T0394857

The
Secret World
of
Denisovans

Also by Silvana Condemi and François Savatier

A Pocket History of Human Evolution:
How We Became Sapiens

The
Secret World
of
Denisovans

The Epic Story of the Ancient Cousins to Sapiens and Neanderthals

Silvana Condemi and
François Savatier

Translated by Holly James
Illustrations by Benoit Clarys

THE EXPERIMENT
NEW YORK

THE SECRET WORLD OF DENISOVANS: *The Epic Story of the Ancient Cousins to Sapiens and Neanderthals*
Copyright © 2024 by Silvana Condemi and François Savatier
Translation copyright © 2025 by The Experiment, LLC
Illustrations copyright © 2024 by Benoit Clarys

Originally published in France as *L'Enigme Denisova* by Albin Michel in 2024.
First published in English in North America by The Experiment, LLC, in 2025.

The Experiment, LLC
220 East 23rd Street, Suite 600
New York, NY 10010-4658
theexperimentpublishing.com

THE EXPERIMENT and its colophon are registered trademarks of The Experiment, LLC. Many of the designations used by manufacturers and sellers to distinguish their products are claimed as trademarks. Where those designations appear in this book and The Experiment was aware of a trademark claim, the designations have been capitalized.

The Experiment's books are available at special discounts when purchased in bulk for premiums and sales promotions as well as for fundraising or educational use. For details, contact us at info@theexperimentpublishing.com.

Library of Congress Cataloging-in-Publication Data available upon request

ISBN 979-8-89303-070-9
Ebook ISBN 979-8-89303-071-6

Jacket and text design by Beth Bugler
Jacket images by Tiia Monto, via Wikimedia Commons, © 2024 Elsevier B.V., AidaTiara – stock.adobe.com (front), Benoit Clarys (back)
Author photographs by Pauline Alioua-Flammarion (Silvana Condemi) and Ingrid Leroy (François Savatier)
Translation by Holly James

Manufactured in the United States of America

First printing August 2025
10 9 8 7 6 5 4 3 2 1

Nothing in biology makes sense except in the light of evolution.

—Theodosius Dobzhansky

Contents

Prologue: The Secret
by Silvana Condemi

I've seen a lot of things in paleoanthropology. I've seen Neanderthal in all its many forms. I've seen and participated in the vast investigation to prove that our fossilized ancestor was European. I've seen paleogenetics barge its way into prehistory. I saw what happened when a micro-snippet of mitochondrial DNA emerged in 1997, suggesting that our ancestors had never mixed with Neanderthals. I saw the entire scientific community wait ten years for this essential piece of anthropological information to be confirmed. And I witnessed the shock when in 2010, the opposite was confirmed to be true: Neanderthals were part of the family. That same year, this surprise gave way to a revelation of an even more chilling nature from Siberia. In a cave tucked away in the Altai Mountains—the Denisova Cave—an unknown human genome had been discovered that was neither Neanderthal, nor Sapiens, nor Martian for that matter.

An unknown species? Surely not. There were no fossils associated with the DNA. None, except for a single fingertip. I'd spent my entire career studying precise characteristics of all kinds of fossils—at least, those large enough to be considered significant.

I'd compared the respective anatomies of fossil bones collected from various species using calipers and cephalometers before the arrival of scanners, high-throughput sequencers, and accelerator mass spectrometry dating. Compared with physics or astronomy, which are thousands of years old, my discipline of paleontology may feel like a young science at a mere 170 years old, but it's one that's been picking up pace at lightning speed. Modern paleoanthropology is now spectacularly well-equipped and has become capable of unlocking the secrets of fossils, to the point of revealing the blood groups of our ancestors, their kinship links, their diets, microbes, migratory trajectories, interbreeding practices, and even their calendar age—all to an outstanding degree of precision. The list goes on. In this respect, the study of Neanderthals is an emblematic one indeed.

In short, I was left stunned by this new discovery. The tiny epiphysis (tip) of the distal phalanx of a little finger, which later appeared to be that of a female adolescent, did not tell us anything whatsoever about her species. I've learned to identify a Neanderthal from a fragment of inner ear, neck . . . practically any part of the skeleton. But this was impossible. Instead, we had to take the word of the new kids on the block, the paleogeneticists, whose techniques are so suspiciously complicated that no one would have been surprised had they announced that Genghis Khan was in fact a Neanderthal, based on his DNA. I shared this sense of unease with other colleagues.

Our training had led us to become accustomed to prehistoric science and its slow, sure pace. Discoveries of new human fossil forms come one after the other, each followed by meticulous work that leaves nothing to chance, often continuing for several years, even decades after initial excavations. Now, these geneticists with their science-fiction methods were making two major breakthroughs per year.

Did my discipline make sense anymore, if nothing could be ascertained without DNA to provide conclusive results? And besides, what were the implications of this discovery of an unknown human in Asia? The Denisova human was nothing short of a mystery.

Journalists, unburdened by convention, had come up with their own solution within a week. They were quick to dub this population, known only by a genome and the fragment of a finger, "The Denisovans." This bold move constituted a huge step forward—and little by little, we paleoanthropologists would have to follow suit. Were we ready to join them? Without noticing it, we had already inadvertently accepted that a "paleontological species" named Denisovans existed. Or rather, must have existed. There was a way to prove it no doubt, but how? We didn't know. With DNA? A human species had never before been defined by its genome alone. And yet we seemed to be accepting this as the first. What did it mean? Was the Denisovan human a kind of Siberian yeti, an abominable snowman that had managed to survive in hiding, much like the diminutive species known as *Homo floresiensis* that inhabited the island of Flores? Or perhaps it belonged to some vast species that had gone unnoticed but nevertheless played an important role in the history of the *Homo* genus in Asia? One of my first thoughts on the subject was that by some miracle, we might have discovered the DNA of the *Homo erectus* (which at the time, was considered a synonym for "prehistoric Asian human").

Back then, it seemed to me that journalists had committed a sin by inventing a pseudo-species and naming it "Denisovan." I expressed my paleontological concerns to François, my coauthor, and himself a journalist. Discussions ensued. Ever since we met, we've been chatting incessantly about prehistory (we've written two books together: one on Neanderthals and the other on Sapiens). We exchange our views on every publication that sheds new light on our planet's ancient human population. We grope for answers, we question scientific paradigms, we look for ways in which to interpret unprecedented findings that prehistorians, paleoanthropologists, and geneticists go to great length to obtain. These exchanges are very precious to me. I still exercise the extreme caution my research environment demands, but I'm also liberated by the attitude of a (cautious) science journalist. I feel free to consider daring hypotheses, which François responds to without the usual reflexes of my colleagues. He in turn proposes his own ideas. The result: We make progress. A lot of progress.

This book is the result of four years of these investigations, and we're finally ready to bring to you the conclusions of our great scientific investigation into the enigmatic Denisovan human. Throughout this book, you'll learn about the evolutionary process in eastern Eurasia that gave rise to the Denisovans. We call this process "Denisovation" in allusion to the parallel process that occurred in western Eurasia known as "Neanderthalization," which gave rise to Neanderthals. Just as Neanderthal once existed in the Far West, in the Far East, there was Denisova. We'll see that prehistoric Eurasians were cultural beings, adapting to every environment they arrived in. They migrated, mixing with the locals they encountered, exchanging ideas and even genes. All of these behaviors turn our current understanding of Asian fossils on its head. But there's no doubt about it: Remarkable discoveries will continue to be made. They will tell the story of how Asia came to be populated and corroborate the findings on Denisovation we outline in this book. In the meantime, the following is the most parsimonious solution possible to the Denisova enigma.

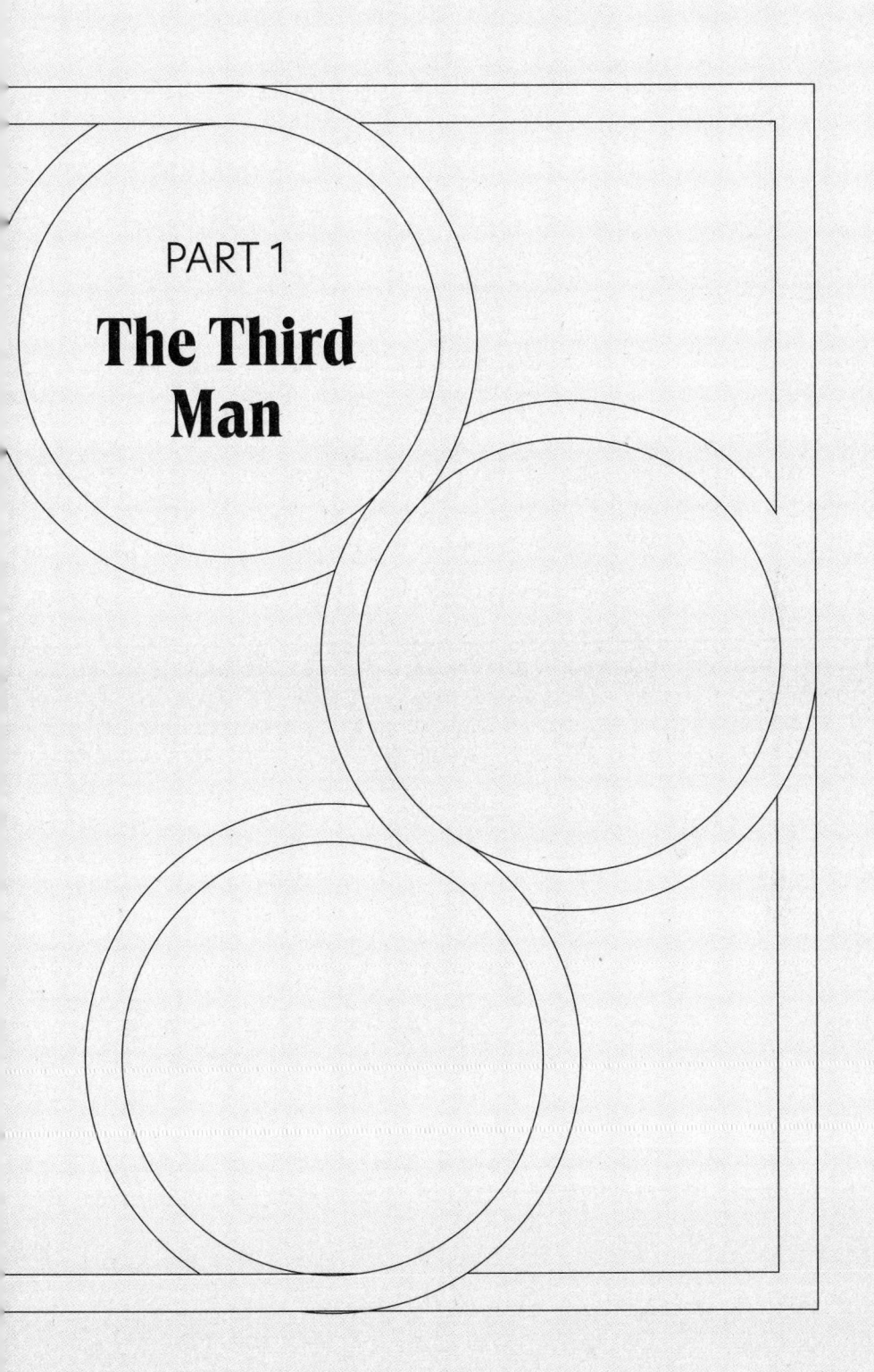

PART 1

The Third Man

This Denisovan hunter from the north bears
a close resemblance to a Neanderthal.

1

A Secret Comes to Light

It's the summer of 2010. The first whistle of the World Cup sounds in South Africa. "This time for Africa!" sings Shakira. It's a fitting statement for soccer fans, but not for prehistorians. By a curious paradox, while the eyes of the entire world are riveted on the cradle of humanity, prehistorians are looking toward Asia—more specifically, to a cave tucked into the side of a valley in an isolated corner of Siberia: the Denisova Cave. It would appear that a never-before-seen human form from the Paleolithic period has reared its head there, or rather, the tip of its finger. The discovery has been made by researchers at the Max Planck Institute for Evolutionary Anthropology (MPI-EVA) in Leipzig.

The Finger that Cast a Chill

For prehistorians, the discovery of a new human species is a rare event. Ordinarily, this kind of occurrence rouses a wave of excitement among the community, alongside innumerable questions: How are these humans from the distant past related to us? Are they our ancestors? Or do they belong to one of the many other branches in the tree of human evolution? Findings of this kind are enough to send shockwaves throughout the global community.

But not this time. Instead, the Leipzig researchers' announcement caused a certain exasperation and deep skepticism. It would turn their world upside down. What exactly was it making the paleoanthropologists bristle? The explanation lies in one other small detail: The researchers who made the discovery were geneticists. Their conclusion was not drawn from the finger fragment they discovered itself, but the DNA they had managed to extract from it. The team shocked the world by publishing almost simultaneously Neanderthal DNA and demonstrating that Eurasians inherited Neanderthal genes. Was it really possible that a new Eurasian species had been discovered? And could this information be determined from DNA alone, from a single fossilized fingertip? Prehistorians were stunned and dismayed about where the geneticists might be leading them.

Wave upon Wave

Strange though it may seem, paleoanthropologists had every reason to believe in this discovery of a new species. This is for the simple reason that it filled a glaring gap in the jigsaw puzzle of human history. To understand this gap, we have to remember that for two million years or more, wave upon wave of humans were making their way from Africa to Eurasia. As the cradle of humanity continued to spill over, new populations went on to meet the descendants of previous waves. When the main wave of Sapiens left Africa some seventy thousand years ago, a species already well established in Europe for hundreds of thousands of years was there to meet them: the Neanderthals.

The cold-hardy *Homo neanderthalensis* made it through four ice ages in Europe, but during the warmer periods members of the species also migrated to the Near East and central Asia. Yet they hadn't always existed. They too were descended from archaic humans, who left Africa in an earlier migratory wave and had spread across the Eurasian continent, where they evolved physically over time. On the European peninsula, these ancestors gave rise to the Neanderthals. That's how, when Sapiens left the African continent, they ended up encountering this robust and resilient human.

We know what happened in the west of the great Eurasian continent. But what about the Far East? What species did Sapiens cross paths with in Asia? If the previous wave gave rise to the Neanderthals in Europe, did it also produce a distinct form of human in Asia? Oddly enough, in 2010, paleoanthropologists were not asking themselves this question, despite the fact that human fossils had been discovered in Asia.

These fossils were attributed to one of the great conquerors of the human family tree: *Homo erectus*. This species of human had a large body and the ability to walk and run, but still had a limited cranial capacity. *Homo erectus* first appeared in Africa and began to disperse from the cradle of humanity after two million years. They are our very distant ancestors, since Sapiens came from the remote descendants of those who remained in Africa.

There were vague assumptions that the Asian population originated from *Homo erectus* in a new, peculiar Asian species. since fossil traces seemed to indicate their presence across the whole of Asia for over a million years. In other words, prehistorians in 2010 believed that "archaic" *Homo erectus* had been succeeded by "evolved" *Homo erectus*—in a new, peculiar Asian species—in a slow and continuous process.

This hypothesis introduced the notion of a dissymmetry between eastern and western Eurasia. In Europe, archaic *Homo erectus* had not evolved into evolved *Homo erectus* but had instead given way to a different species entirely: Neanderthal. So why, when it came to Asia, did we favor the astonishing, and rather implausible, hypothesis of continuous evolution having taken place over an extremely long period of time (a third of the lifetime of the *Homo* genus) rather than consider the possibility of the emergence of a new species?

But despite the double standard that arises from this perspective, paleoanthropologists remained adamant: Before Sapiens, *Homo erectus* had been the sole inhabitant of the immense Asian continent, save for a few incursions by Neanderthals. Which is why the discovery of a new species in the Denisova Cave caused brows to furrow—to say the least—throughout the paleontology community.

The second reason, as we've seen, was that the revelation came from a genetics laboratory. It was a Swedish biologist by the name of Svante Pääbo who oversaw both the research on the phalanx fragment unearthed in the Denisova Cave and the research on Neanderthal DNA that had proved so disturbing for the community of prehistorians. This visionary and founder of paleogenetics was awarded the Nobel Prize in Physiology for Medicine in 2022 in recognition of his contribution to scientific progress.

Svante Pääbo, winner of the 2022 Nobel Prize in Physiology and Medicine, holding a cast of a Neanderthal skull in his hands

Svantastic

Svante Pääbo was the first visionary to consider the possibility of finding ancient DNA in fossils. With his fighting spirit, he encountered and ultimately overcame the greatest obstacle in his path: contamination. His name often appears at the end of the list of authors when major paleogenetic discoveries are published. Those familiar with how research works will know that when it comes to scientific publications, the first and last names on the list of authors are the most significant, the first being the person who carried out the work and the last being that of the research director who initiated, devised, and managed it. Journalists have grown to recognize Svante Pääbo's name as the sign of a major breakthrough. While his invaluable contribution is celebrated today across the world, Svante's career began much like many other scientific sagas, with the young scientist working alone and with major technical difficulties to overcome.

While he was completing his PhD in the 1980s, Svante Pääbo was practically the only one to believe it would ever be possible to extract and analyze ancient DNA. At the time, even hoping was considered madness; the prevailing opinion was that any genetic material more than a few years old was lost forever. And for good reason. DNA degrades very quickly, not least due to the tens of thousands of lesions each cell is subjected to each day as a side effect of natural radiation on Earth's surface. These lesions are constantly repaired when an organism is still alive, but not after death. Necrophagous organisms (those which consume decomposing biomass) spring into action after a living being dies, leaving destruction in their wake

During his spare time, Svante Pääbo secretly conducted tests on DNA taken from mummies in the Egyptological collection of the Uppsala Museum in Sweden. He gathered all the pieces of DNA collected from each mummy and attempted to put them together like a puzzle. At the time, the only technique used for DNA restitution was what is known as "polymerase chain reaction," the good old PCR used for the COVID tests we've become so familiar with. But the efficacy of this method was limited by the contamination of the sample by everyone who had touched it, be they embalmers in ancient Egypt, archaeologists, museums employees, etc.

In 1984, Svante Pääbo succeeded in isolating genetic material from cells extracted from mummies for the first time. After this initial success, he dedicated his energy to finding ways to overcome the issue of contamination of ancient DNA with contemporary DNA. We may not be aware of it, but we are constantly releasing our DNA into the environment. Svante Pääbo invented

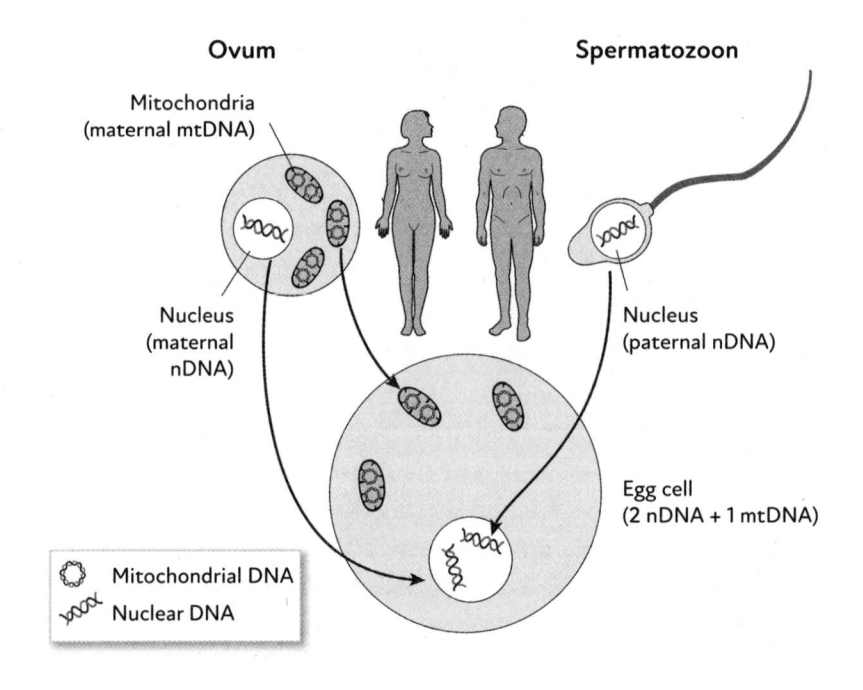

Just like plants and animals, humans possess two types of DNA: nuclear DNA (nDNA), transmitted by both parents, and mitochondrial DNA (mtDNA), transmitted only within female lineages. While the spermatozoon does contain mitochondria, these are not transmitted to the egg cell from which the child develops. The male sex (Y) chromosome, on the other hand, is only transmitted within the male lineage in mammals.

methods for recognizing and chemically isolating fragments of DNA from a specific organism so that they can be combined using bioinformatics.

In 1990, Ludwig Maximilian University in Munich had the foresight to recruit this brilliant researcher, then just thirty-five years old. Thanks to the techniques he had developed, Svante Pääbo was able to extract uncontaminated DNA from a Neanderthal arm and use it reconstruct a 379-base sequence.

These bases are contained within nucleotides, the molecules that make up the DNA, and form part of a 16,569-base ring that constitutes the mitochondrial DNA of the Neanderthals discovered in the Neandertal valley. The progress made by Svante Pääbo's team was so promising that in 1997, the German Research Foundation created the Max Planck Institute for Evolutionary Anthropology in Leipzig, which was endowed with substantial technical, human, and financial resources, and appointed Pääbo director of the genetics department. The Max Planck Institute for Evolutionary Anthropology will come up numerous times throughout this text because it played such a major role in paleoanthropology, so we'll shorten its name to MPI-EVA.

At the turn of the millennium, Svante Pääbo and his team set out on the path that would one day lead them to drop a scientific bombshell, thanks to the methods they'd developed: They began the process of sequencing the complete genome of the Neanderthal species. In 2006, Silvana co-organized a major conference in Bonn, not far from the Neandertal valley, where the first fossil of the eponymous species was discovered in 1856. Prehistorians came to commemorate the 150th anniversary of this discovery, hoping for a presentation of the latest updates on Neanderthal genetics—and perhaps even the complete sequencing of the mitochondrial DNA. But determining Neanderthal's nuclear DNA still seemed light years away.

This is because humans possess two genomes. One, the mitochondrial DNA, is contained in the few hundred to several thousand mitochondria of our cells, the tiny cellular organelles that produce most of the cell's energy. The other, nuclear DNA, is contained in a single copy in the cell nuclei. The system of 3.2 billion base pairs contained in the nuclear DNA is formed with each new generation from a combination of each parent's nuclear DNA. Meanwhile, the 16,569-base ring of the mitochondrial DNA is contained in the mitochondria of the oocyte, meaning it is transmitted from mother to child unchanged. Studying the latter form of DNA therefore only allows us to trace female lineages, while the study of the Y chromosome allows us to trace male lineages.

A Nuclear Bombshell

This is why in 2006, the sequencing of Neanderthal's mitochondrial DNA—let alone its nuclear DNA—seemed so out of reach. But lo and behold, during the congress, Svante Pääbo announced that his laboratory was already working on sequencing the Neanderthal genome in its entirety. Given that the complete sequencing of the *Homo sapiens* genome had only been completed in 2004 after fifteen years of an enormous international effort and three billion dollars in financing, everyone present was stunned by Svante Pääbo's announcement. The methods the researchers had invented were so efficient, it took just two years to obtain the complete mitochondrial DNA of one Neanderthal. In 2009, mitochondrial DNA from five further Neanderthals followed. Each one appeared to confirm the paleoanthropologists' hypothesis that *Homo sapiens* and *Homo neanderthalensis* were two distinct species that could not interbreed.

But this hypothesis proved to be false. The sequencers had become so productive that after just two years, on May 7, 2010, the MPI-EVA team and David Reich's team at Harvard University jointly published 60 percent of Neanderthal's nuclear genome in the journal *Science*. The genome in question was a sort of Neanderthal chimera obtained by combining DNA remnants from three bone fragments found in Vindija Cave in Croatia.

The nuclear DNA was expected to confirm that *Homo neanderthalensis* and *Homo sapiens* could not interbreed, but in fact it revealed the opposite. Professor Richard E. Green announced the curious findings thus: "We are now able to say that, in all likelihood, gene transfer has occurred between Neanderthals and humans."

Humans. It's worth spending a moment on terminology here. In his statement, Reich seems to be creating a distinction between Neanderthals and "humans." In the anglophone world, the term "human" tends to be reserved for *Homo sapiens.* But it can also be used to designate the *Homo* genus in all its many forms.

The announcement in 2010 had the potential to change all the hypotheses that had been made from fossil discoveries thus far. Contrary to what had been thought previously, Sapiens and Neanderthal may have in fact interbred. Their lineages were mixed. Paleoanthropologists estimate that the genomes of all Eurasians today contain between 1 and 4 percent Neanderthal DNA. Eurasian is used here not to describe a person with one European and one Asian parent, but simply any inhabitant of Eurasia, the supercontinent made up of Europe and Asia.

Thanks to this and subsequent research, we now know with certainty that modern-day Eurasians have several ancient human ancestors, including the Neanderthals. This suggests that members of the main wave of Sapiens from Africa around seventy thousand years ago not only encountered Neanderthals along the way but also interbred with them, thus acquiring Neanderthal genes and retaining only the most useful of them. It is now estimated that the Eurasian genome as a whole contains no less than 30 percent of the Neanderthal genome, which is a lot.

In 2010, this information was still unknown. Paleoanthropologists of the time remained prisoner to their traditional working methods, which caused them to look askance at these disconcerting findings. For 170 years, they had been accumulating biological knowledge about ancient humans from bones and teeth. Back in 2010, they had barely noticed that there were now geneticists working in the field of prehistory or found their obscure techniques difficult to understand. And difficult to accept.

In fact, prehistorians hadn't yet grasped the enormity of what was happening. Thanks to developments in the field of industrial DNA sequencing, and the computing power that had become available in bioinformatics, it was now possible to study the genomes of Paleolithic populations and acquire biological information about them. Prehistorians continued to turn a blind eye when, on December 22, 2010, the same team presented a previously unpublished nuclear DNA extracted from a human, from a phalanx fragment found in a cave in the Altai Mountains in Russia, a finding that was to prove extremely significant.

Paleontologists Have a Bone to Pick

The tiny fragment of bone appeared to have belonged to a young person, as it had not yet been fused to the rest of the phalanx. Measuring around one centimeter square, it had yielded enough genetic material to make it possible for researchers to reconstruct both the mitochondrial and nuclear DNA sequences and determine the individual's sex. Svante Pääbo and David Reich specified that the adolescent in question was not descended from *Homo erectus*. This discovery was all the more astounding given that the DNA sequencing had been developed from a bone that had too few distinctive features for paleoanthropologists to even link it to a known species.

This was an unprecedented occurrence in paleoanthropology. For the first time, a new human form—Denisova—had been identified, not from fossils but from its genes. As soon as Denisova's mitochondrial DNA was published, Svante Pääbo himself described being "extremely surprised" and cautioned it would be necessary to wait for the sequencing of the nuclear

DNA before researchers could be sure they were dealing with was a new human species. But by December 2010, the wait was already over, and Denisova's nuclear DNA was published, making a mark in the prehistoric world.

Was the discovery to be believed? Paleogenetics was a nascent discipline, yet its pioneers were already claiming to have uncovered a new species thanks to their advanced expertise in chemistry, biology, genetics, statistical mathematics, and computer science. But the paleogeneticists' announcement had been made, and the rest of the scientific community would have to proceed accordingly—i.e., either refute the discovery or confirm it.

The site of the discovery, the Denisova Cave, was then known only to a small number of prehistorians working in central Asia. These experts who had made remarkable discoveries in prehistoric science in the Soviet Union were already familiar with the Altai Mountains and were aware that Neanderthal fossils had been discovered there in caves around 60 miles (100 km) from Denisova. But the question was, did the Denisova adolescent belong to a new species, or were researchers dealing with a new variety of Neanderthal? And if this really was a new species, was it possible that it had crossed paths with Neanderthals? Had they met the "Denisovans" in the flesh?

The first reaction after the discovery was made was one of doubt. Many respected paleoanthropologists urged great caution in light of the paleogeneticists' conclusions. Michel Brunet, the researcher who discovered a prehuman biped known as *Sahelanthropus* (nicknamed "Toumaï") dating back seven million years, declared: "It is of utmost importance that we take our time. If these were colleagues in my own research team I would send them out into the field immediately to try and find new bones to confirm these interesting findings. That is not to say I do not trust this work. Svante Pääbo is, without a doubt, one of the world's leading specialists in his field. But until further discoveries are made, it's probably best to remain cautious."

Perhaps the paleoanthropologists' spectacular success in applying their methods to fossils had gone to their heads and they were now going overboard? Who were these so-called Denisovans with little to no fossil material? Had they existed at all?

A Sibling from Asia

The Denisova findings constituted a scientific tour de force. But remarkable as this breakthrough may have seemed, its real interest lay elsewhere. With the help of their powerful computers, researchers were able to calculate the most likely kinship tree linking the Sapiens, Neanderthal, and Denisovan mitochondrial DNA. The results suggested that the ancestors of all three species—rather than originating directly from *Homo erectus*—separated paths around one million years ago. It may have been too early for paleoanthropologists to draw meaningful conclusions from this date, but it suggested that Denisova was a kind of "Asian Neanderthal," placing Denisova on the map of Eurasia in the same way Neanderthals had been traced back to Europe.

If Denisovans didn't descend from *Homo erectus*, where did they come from? Africa? From an isolated population in Asia that had undergone genetic drift over a long period of time? These questions seemed all the more enigmatic given that the December 22, 2010, publication also revealed that the nuclear genomes of certain Asians contained Denisovan DNA.

In 2010, making sense of this information felt like an impossible task, even though there was a sense among prehistorians that several major discoveries were being made one after the other. It was now known that two different human forms had coexisted in Eurasia during the Paleolithic period: the European Neanderthal and the Asian Denisovan, whose existence had been genetically proven but remained elusive for paleoanthropologists.

Today, over a decade has passed, and prehistorians have gone from shock to acceptance of this new human from Asia. This collective acknowledgment has inevitably been accompanied by a whole host of new questions. What did the Denisovans look like? What kind of lives did they lead? How did they manage to adapt to cold regions? What legacy did they leave to modern Asian populations? And has this legacy been beneficial or disadvantageous?

Though the mystery surrounding the Denisovans has not been completely resolved, we believe we have now cleared up most of the gray areas. To understand the true nature of this ancient human being from Asia, we have to go back to the cave where it all began.

A cozy mountain apartment for humans, the Denisova Cave
was also a much-frequented den for hyenas.

2

In the Denisova Cave

Siberia. Just over 90 miles (145 km) from the mining town of Barnaul, a river meanders through meadows and pine forests, creating a landscape reminiscent of the Alps. But this river is winding its way through the Altai Mountain range. A crossroads of peoples and languages since time immemorial, this mountain range stretches around 1,200 miles (1,931 km) from Kazakhstan to the Altai Republic in Russia, Mongolia, and the Xinjiang Uygur Autonomous Region of China. Here, the mountain range tapers off into the verdant hills of the valley of Sibiryachikha. Along a river bend, a cavity appears in the cliff face.

The cave is known as the "Bear's rock" by the Altai people, Turkophones who have long inhabited the region. The Russians, on the other hand, have dubbed the cave Denísova peshchéra, meaning "Denis's cave." According to local tradition, an "old-believer" (i.e., someone who refused to accept the reform of the Orthodox Church promulgated in the mid-seventeenth century) who went by this name once lived here as a hermit. It's a great story, but it's also possible that the name comes from Denisous, a shepherd who made the cave his sheepfold.

Prehistoric humans are thought to have made the site their home because prehistoric people traveled along valleys where rivers attract wildlife and provide pebbles for shaping tools. By Paleolithic standards, they hit the luxury mountain apartment jackpot when they stumbled upon the Denisova Cave. The high ceilings of its three spacious rooms cover a total surface area of almost 3,000 square feet (279 sq m). It would have been enough to comfortably house the whole clan—once the hyena squatting on the premises had been evicted.

It was in the soil of one of these rooms that the tiny piece of bone that held the key to a phantom species was unearthed. Despite the initial absence of conspicuous fossil material, the site has yielded vital information about the Denisovan people, their daily lives, and the Altaian geographical and climatic context in which they lived. But it was not yet clear whether the river alone could explain why the Denisovans lived there. Nor was it clear whether they lived in this rock shelter alone, or whether they crossed paths with other prehistoric humans. Luckily, the conditions inside the cave where excavations took place provide a wealth of information to help us answer these questions and others.

The Altai: An Eternal Crossroads

The average temperature in the cave, which sits at an altitude of almost 2,300 feet (701 m), is 32°F (0°C)—a little on the cold side for a luxury penthouse. But for the prehistoric peoples who lived there, it was a comfortable temperature in comparison to the cold temperatures outside in winter. During the interglacial periods ("warm" intervals between two glacial peaks) these humans would have endured a continental climate similar to today's, marked by temperatures of −4°F (−20°C) in winter and peaking at 77°F (25°C) at the height of summer.

The presence of prehistoric hunter-gatherers in these mountains may seem unexpected given this harsh climate. In the Altai, the only humans who can survive are those who know how to protect themselves from the cold. They not only needed shoes to be able to walk in the snow, they also had to master fire, dress themselves, or alternatively, coat their bodies with animal fat to resist the cold and humidity, not to mention find caves to take refuge in during the winter. Denisovans mastered these skills, just like their Eurasian contemporaries did.

Indeed, the Altai was a kind of crossroads where different peoples would have encountered one another since time immemorial. The mountain range is home to numerous caves, which Neanderthal clans inhabited in winter, often spending the summer in the open air. The Ust'-Kanskaya Cave and the Kara-Bom open-air archaeological site are around 37 and 62 miles (50 and 100 km) from Denisova, respectively. The Chagyrskaya and Okladnikov caves, 43 and 62 miles (69 and 100 km) from Denisova, respectively, were occupied between sixty thousand and forty-four thousand years ago and are littered with flaked stones and Neanderthal fossils. Incidentally, the Okladnikov cave is named after the great Russian prehistorian, Alexey Okladnikov (1908–1981), a pioneer of archaeology in the Altai (see the photo on page 27). In the Ust'-Kanskaya Cave (also known as Ust-Kansky) in the valley of the Charysh River, the discovery of several bovine and equine carcasses also suggests the presence of Neanderthals in Siberia around fifty thousand years ago.

Yet another population is also known to have visited central Altai: our own. The Kara-Bom archaeological site, 62 miles (100 km) from the Denisova Cave, is one of the oldest sites attesting to the arrival in the region of Upper Paleolithic industries, and therefore, a priori, *Homo sapiens*. The oldest evidence of the presence of our species found in the high latitudes (the central segment of a femur) came from a place called Ust'Ishim, over 600 miles (965 km) to the northwest of the Denisova Cave on the banks of the Irtysh River.

It has also been possible to determine the hunting ground of these humans. The distribution of prehistoric sites in the Altai shows that hunter-gatherers moved over great distances along rivers. While the earliest prehistoric Altai populations were somewhat less mobile, the frequency of travel increased over time, particularly from around forty-three thousand years ago with Sapiens. Certain sites seem to have been conducive to certain activities, such as toolmaking, butchering carcasses, etc.—activities that Sapiens, Neanderthal, and Denisovan hunter–gatherers all participated in.

In the Denisova Cave, the stone used to carve tools came from surrounding streams. The mediocre quality of the stone illustrates the opportunistic behavior of hunter-gatherers. They would seek out territories with an abundance of game, even if the raw material supply left much to be desired. Even so, the river network that enabled humans (including Denisovans) to move around more easily guaranteed a supply of lithic materials even in snowy conditions.

The Curse of the Hungry Hyenas

Evidence of these activities suggests a very early presence of humans in this region of northern Eurasia. Soviet prehistorians Alexey Okladnikov and Anatoly Derevyanko of Novosibirsk State University had already pointed this out, but their colleagues in the West remained in the dark due to their inability to read Russian. Though studies on the Denisova Cave began in the 1970s, the first excavation reports did not emerge from the USSR until ten years later. These reports describe tens of thousands of lithic tools and bone fragments, most of which had been chewed up and digested by hyenas.

The fact that these sites were frequented by hyenas may seem like a mere detail, but it plays a crucial role. When female hyenas are pregnant, they need a quiet place to take shelter and chew on bones to obtain the calcium they need for the embryo to grow. As a result of being chewed up by hyenas and continually trodden on by the cave's inhabitants, the tens of thousands of fossils found on site consist mainly of bone fragments no larger than a

In 1971, archaeologists from the northern Asia met in Ulaanbaatar,
the capital of Mongolia. The great Russian prehistorian Alexey Okladnikov
(1908–1981), who spent decades working on excavations in the Altai, is pictured
second from the right. Anatoly Derevyanko, whose work is mentioned several times
in this book, is pictured on the far left, wearing a beret.

centimeter in size. What's more, the hyenas' digestive juices left
them barely recognizable, making it impossible to attribute them
to any species. Those pesky hyenas . . .

It was clear that the cave had been inhabited long ago, but in
the absence of well-preserved bones, it was impossible for prehistorians to determine whether these occupants were Neanderthals,
Sapiens, or another species entirely. All researchers had was an
idea of the general chronology of when the region had been
populated.

Like the details of the Soviet excavations in the Altai, the dates
of occupation of the region were slow to reach the Western
community of prehistorians. The lack of fossil material and the
fact that early publications were in Russian only hindered the

dissemination of these very significant discoveries coming from Denisova. It wasn't until 1999, after the fall of the Iron Curtain, that Stanislav Arkhipov of the Unified Institute of Geology and Minerology in Novosibirsk shared with non-Russian-speakers the likely arrival dates of the first Siberians. Unsurprisingly, it would appear that humans only migrated to the region during the warmer interglacial periods.

Stanislav Arkhipov recalled that Anatoly Derevyanko had determined a first sporadic human settlement of the Altai (possibly *Homo erectus*), as indicated by flaked stones dating back some 800,000 years. He then reported that he had recorded five successive occupations, each corresponding to interglacial periods ranging from 424,000 to 374,000 years ago, 337,000 to 300,000 years ago, 243,000 to 191,000 years ago, 123,000 to 109,000 years ago, and finally, 100,000 to 50,000 years ago, when our own species arrived.

DNA Fridge

During these hundreds of thousands of years of successive occupations, the average temperature in Denisova remained close to 32°F (0°C), so the DNA contained in the bone fragments was preserved. You might be wondering what exactly is meant by "preserved." Even in bone, DNA degrades very easily, and what remains is drowned in genetic material from bacteria, fungi, and all the soil organisms that feed on the body of a living being after it dies.

While discoveries of mitochondrial DNA are much more common, nuclear DNA can only be recovered if it has been preserved under exceptional conditions, for example the kind of cold temperatures found in Denisova. The cave was therefore of particular paleogenetic interest for both Russian prehistorians in Novosibirsk and paleogeneticists at MPI-EVA in Leipzig, who were always on the lookout for fossils likely to contain DNA. In 2008, Anatoly Derevyanko entrusted Svante Pääbo with a fragment of bone from the Denisova Cave.

Now estimated to be between fifty-two thousand and seventy-six thousand years old, the fragment—part of the epiphysis of the first phalanx of a little finger—measures 5 by 7 millimeters and weighs just 40 grams. According to standard procedure, it was named *Denisova 3*: the name of the site and a number used to identify a fossil (or set of fossils) belonging to the same individual. Thus, "Denisova" does not designate the species the specimen belongs to, which could just as well be Neanderthal. Another human bone was found at the same time: the diaphysis (the long part of a bone) measuring 15 by 7 millimeters, belonging to the same phalanx.

Russian prehistorians had already determined that these were human bone fragments belonging to a relatively young individual. They knew this because in modern humans, the phalanges do not fuse until the age of sixteen. Given that adolescence corresponds to a much earlier age in prehistoric humans, the age of death of *Denisova 3* has been estimated at twelve to thirteen years.

In his book, *Neanderthal Man: In Search of Lost Genomes*, Svante Pääbo explains that *Denisova 3* was so small that there was little hope that the specimen would contain enough DNA. "I didn't get around to analyzing the sample for six months," he writes. So low was the measly bit of finger on his priority list of fossils to study.

But it turned out that the little bone contained enough DNA to reconstitute an entire genome. Researchers had no choice but to accept what *Denisova 3* revealed to them: A third human form had populated Eurasia in the Paleolithic period at the same time as *Homo neanderthalensis* and *Homo sapiens*.

Interestingly, the excavators decided to separate the two pieces of the *Denisova 3* fossil. The smaller piece was sent to MPI-EVA, while the larger fragment (the diaphysis) was loaned to the University of California, Berkeley. The "large piece" seemed so unpromising that researchers neglected to test it for DNA, and it was forgotten, to the point where it could no longer be identified—what now seems like a colossal oversight in light of the subsequent discovery of the Denisovan species. A Franco-Russian

Excavators at the Denisova Cave have to wear full bodysuits to retrieve samples from this DNA fridge.

team, led by geneticist Eva-Maria Geigl from the Institut Jacques Monod in Paris, set about identifying it by its mitochondrial DNA, which enabled the two *Denisova 3* fragments to be reunited into a virtually complete bone that turned out to resemble a Sapiens phalanx more than a Neanderthal phalanx. And this was not the only part of the Denisovan skeleton that seemed to have more "sapien-esque" than Neanderthal features, as we will see in chapter 12.

A Bone of Contention

As we've already seen, the publication of the MPI-EVA results provoked a general outcry. Prehistorians found it hard to accept a "third Pleistocene human" on the sole basis of DNA. The concept of genetic species did not yet exist; the only known and recognized species were paleontological species, determined by bone shape.

And so, attempts were made to circumvent this dilemma. The geographical location of the Denisova discovery prompted some to suggest an "Asian variety" of Neanderthal. Others argued that the genetic material in question corresponded to the paleontological species identified long ago in China, that of *Homo erectus*. According to proponents of this theory, "evolved" *Homo erectus* had taken refuge in Siberia, "pushed out" by our own Sapiens ancestors who were aggressively taking over Asia having left mainland Africa between eighty thousand and sixty thousand years ago.

The substantial size of the teeth found in the cave seemed to corroborate their theory. The discovery of an enormous upper third molar in 2008 provided further proof that macrodontia was a trait belonging to the individuals in question. The study of this tooth, alongside *Denisova 4* (a second or third upper molar), highlighted their astonishing volumes. All the dimensions of the crowns, roots, contours, etc., were far superior to those previously found in *Homo neanderthalensis* or *Homo sapiens*.

The macrodontia, or even megadontia, observed in these specimens is all the more remarkable given that the occlusal surfaces (surfaces for chewing) have complex morphologies: They each have six or seven cusps (points). This finding suggests the individuals in question had evolved to chew tough plants.

Interestingly, in this regard, they also resemble those of *Homo erectus* from the Sangiran archaeological site in Java. According to work published in 2023, this species also consumed the many nourishing plants available to them (grasses, fruits, etc.) during the wet season, as well as tougher, less nourishing plants in the dry season, which they would have had to spend a long time crushing in order to extract any nutrients.

Denisovan dental characteristics therefore suggest that their population evolved mainly in the tropics just like *Homo erectus*.... This explains why, despite genetic evidence to the contrary, some paleoanthropologists insisted that Denisova was in fact an "evolved" *Homo erectus*. Their skepticism was natural for experienced researchers in their domain, but the subsequent discovery of other genomes, along

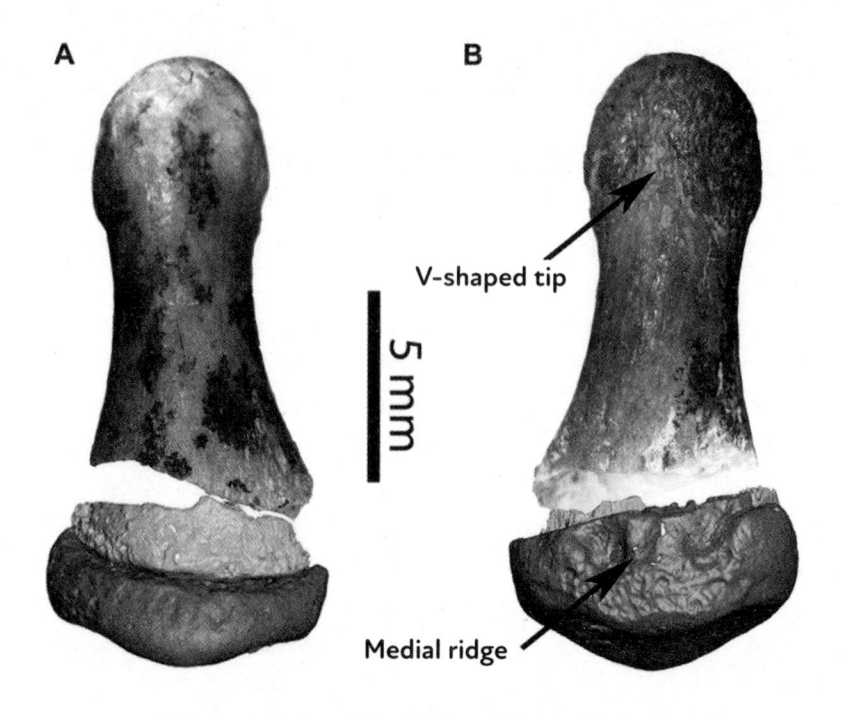

A 3D reconstitution of the *Denisova 3* phalanx: The upper part had been sent to UC Berkeley for study; MPI-EVA researchers found the Denisovan DNA in the lower part.

with the fact that Neanderthals and Denisovans lived side by side at the Denisova site eventually constituted evidence that the specimen in question was that of a distinct human form. But *Denisova 3* was not the only surprise to come from the cave.

ZooMSing in on Denisova

Now aware of the immense value of the cave's fossils, paleoanthropologists have been working to better identify, date, and sequence Denisova's DNA. Today, a remarkable method exists that makes it possible to identify humans or animals on ancient bones (Zooarchaeology by Mass Spectrometry) and DNA in sediments. This technology makes it easier to study successive human occupations by looking at the stratigraphy of a cave. The analyses of the Denisova Cave benefited greatly from these advances: because the strata in the three excavated areas of the cave had been disrupted to such an extent that it would have proven extremely difficult without it.

ZooMS technology consists of an extremely powerful screen for identifying the species of a bone fragment and was a crucial development for this excavation. We have already seen the devastating effect of the hyenas' chewing on the cave's fossils. Although over 135,000 bones have been discovered at the Denisova site since 2008, 95 percent of them are too damaged to be identified by the naked eye. *Denisova 5*, the phalanx of a Neanderthal foot unearthed in 2010, is practically the only exception.

Another name for ZooMS might be "collagen peptide fingerprinting." Peptides are polymers that are linked together to form amino acids. Collagens are polypeptide macromolecules that have important structural roles to play in the body. Collagen composition varies slightly from one animal family to the next. By comparing the composition of a certain bone collagen with those recorded in the library of peptide collagen compositions of known animals, it is possible to assign a given bone fragment to a family, a genus, or even a species.

When combined with reliable dating techniques, this method makes it possible to extensively and accurately determine the faunas (indicative of the type of environment in question) and dates of successive human occupations. It was a team led by Katerina Douka—at the Max Planck Institute in Jena at the time—who decided to use ZooMS at the Denisova site. Among the multitude of animal bone fragments studied, the researchers identified twelve fragments of human remains, which they then dated. A few more human bone remains were discovered later in 2022. As the majority were too old for carbon-14 dating, researchers dated the specimens using a probability method known as the Bayesian model to link them to the known and dated genomes of Denisova, Neanderthal, and Sapiens.

The oldest human sedimentary DNA (which is in fact Denisovan) dates back between 374,000 and 337,000 years, whereas the site's oldest fossil is that of a Denisovan who lived in the cave between 194,400 and 122,700 years ago. Later, it turns out that two Neanderthals (*Denisova 5* and *Denisova 9*) were also present between 130,300 and 90,900 years ago. It's worth noting that the oldest sediments containing traces of Neanderthal DNA date back some 190,000 years. *Denisova 11*, which we will come back to, lived between 118, 100, and 79,300 years ago, while the most recent of the Denisovan fossils, the famous *Denisova 3*, was found in a stratigraphic position comparable to those of the two Neanderthal fossils, which means it would have lived between 76,600 and 51,600 years ago. One fossil, *Denisova 14* has been carbon-dated to some 46,300 years ago, but due to a lack of DNA, it has not been possible to link it to Neanderthal, Denisova, or Sapiens. All this information led to a clear conclusion: For hundreds of thousands of years, Denisovans and Neanderthals must have mixed in this cave.

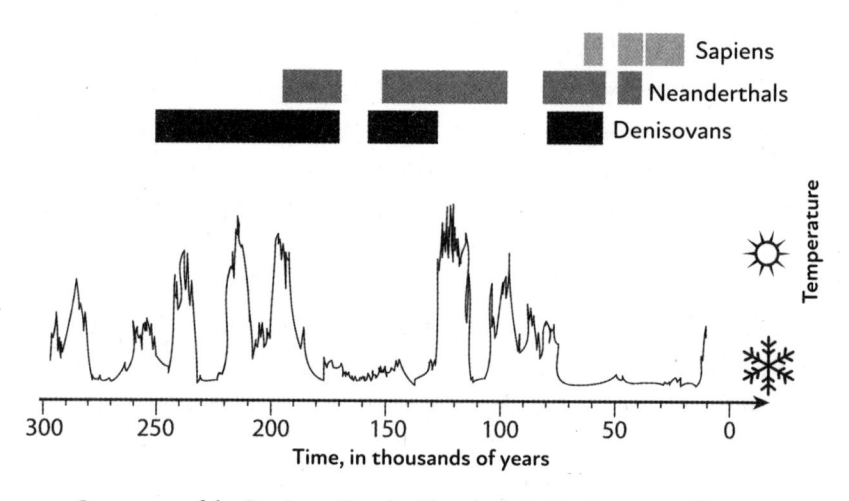

Occupancy of the Denisova Cave by Neanderthals, Denisovans, and Sapiens over the course of the last 300,000 years

Aside from *Denisova 1*, *7*, *10*, and *12*, which have not been published, the human fossils discovered at the Denisova site are the following:

- *Denisova 2*, the very worn second lower milk molar of a Denisovan
- *Denisova 3*, the fifth phalanx of the little finger of a Denisovan
- *Denisova 4*, the second or third upper molar of a Denisovan
- *Denisova 5*, the proximal toe phalanx of a Neanderthal, which has yielded the best Neanderthal genome we have to date (52x)
- *Denisova 6*, the lower incisor of a human that did not contain DNA. Neanderthal? Denisovan? Sapiens?
- *Denisova 8*, the third upper molar of a Denisovan
- *Denisova 9*, the last phalanx of a hand, without DNA, possibly Neanderthal
- *Denisova 11*, a long bone fragment from a Neanderthal/Denisova hybrid
- *Denisova 13*, the fragment of a parietal (skull) bone measuring 8 by 5 centimeters, unpublished
- *Denisova 14*, a fragment of human bone from an undetermined species
- *Denisova 15*, a fragment of a long human bone from an undetermined species
- *Denisova 16*, a mini fragment of bone from an undetermined species
- *Denisova 17*, an unidentifiable fragment attributed to Neanderthal
- *Denisova 18*, a human fragment from an identifiable species
- *Denisova 19*, an unidentifiable fragment attributed to Denisova

- *Denisova 20*, an unidentifiable fragment attributed to Denisova

- *Denisova 21*, an unidentifiable fragment attributed to Denisova

The majority of the fossils in this list are remarkable. *Denisova 5*, a Neanderthal dating back some 130,000 years, yielded the highest-quality Neanderthal genome known to date. Though they contained very little DNA, *Denisova 2, 4*, and *8* have helped determine periods of both Neanderthal and Denisovan occupation in the cave. *Denisova 11*, nicknamed "Denny," is the first known human hybrid in history. The presence of both Denisovans and Neanderthals in the cave is not surprising, since over the past decade Russian prehistorians have identified Neanderthal remains in the Okladnikov and Chagyrskaya caves, which are located not far from Denisova. This has been confirmed by mitochondrial DNA analysis. In the past, this kind of species sorting at Denisova would have been inconceivable; it goes to show what clarification ancient DNA sequencing offers us when it's possible. Without the help of this extraordinary method, the bone fragments uncovered in the Denisova Cave would have likely been classified as Neanderthal, or even been taken for animal specimens.

What were the climate conditions these individuals were living in? In parallel with the work of Katarina Douka's team, Zenobia Jacobs's team at the University of Wollongong, Australia, has been studying the transformations of Denisova's environment over time, on the basis of no fewer than 103 optical dates (determined using OSL or Optically Stimulated Luminescence Dating) ranging from 300,000 to 20,000 years, and the study of the remains of 27 species of large vertebrates, 100 species of small vertebrates (mammals, fish, etc.), and 72 species of plants. The results confirmed in detail what Russian researchers in Novosibirsk had already outlined: The variations between pine, birch, and deciduous forests during warm periods and tundra during cold periods coincide with those established for the environment of Lake

Baikal, 995 miles (1,600 km) farther east, and as expected, human occupations took place during interglacial periods.

History's First Hybrid Hominin

It bears repeating that the discovery of the Denisovan Neanderthal fossils has significantly advanced our understanding of *Homo neanderthalensis*. The quality of the DNA sequencing of *Denisova 5*, a toe phalanx, was remarkable. This work led to several important observations: Firstly, the highly reliable sequencing of this Neanderthal genome made it possible to estimate the percentage of Neanderthal DNA in all non-Africans at around 2 percent on average. Next, it was established that *Denisova 5* was a woman conceived either by a half-brother and half-sister, a pair of double cousins (where the fathers are brothers and mothers are sisters), or—less likely—by a grandfather and his granddaughter, or a grandmother and her grandson. . . . In other words, there was a remarkably high rate of inbreeding among the group of Neanderthals to which *Denisova 5* belonged.

What might this tell us? The survival instinct of these small clans probably led them to interbreed with neighbors. In order to identify possible family links between neighboring clans, a team led by Laurits Skov from MPI-EVA sequenced the DNA of thirteen individuals from the Chagyrskaya and Okladnikov caves in 2022. The result was that by identifying mitochondrial DNA common to both caves, the researchers were able to identify a societal phenomenon already known among Neanderthals: patrilocality. In other words, women would change clans to have children.

A high level of inbreeding was also detected within each group, reflected in a high level of homozygosity (the presence of the same allele—i.e., the same coding DNA sequence, on both the paternal and maternal chromosomes). So much so that the Chagyrskaya Neanderthals had the same level of homozygosity as the mountain gorilla, an endangered species whose six hundred remaining members live in the wild in communities of only four to twenty individuals.

Thus, the Neanderthal and Denisovan populations of the Altai became so scarce that they came to mix, as illustrated by *Denisova 11*, an unidentifiable bone fragment (perhaps a femur or humerus), which ZooMS revealed to be human. As it turned out, this fossil contained well-preserved DNA, which Viviane Slon's team at MPI-EVA was able to sequence.

The complete genome turned out to be that of a young girl around thirteen years of age. The researchers found that the distribution of genes in this genome could only be explained by the individual having a Neanderthal mother and a Denisovan father. *Denisova 11*, or "Denny" as the researchers affectionately named her, is therefore the first and only first-generation Paleolithic hybrid ever discovered. The first hybrid hominin in the history of humankind. Another remarkable discovery in the new science of paleogenetics.

This gives us a clear overall picture of human life in the Denisova Cave: For tens of thousands of years, during the interglacial ages, Neanderthals and Denisovans—two human forms closer to each other than they are to *Homo sapiens*—frequented the cave and met there. Some researchers even believe that these two species may have founded a common culture in the Altai. Over the last three hundred thousand years, this region has always been inhabited by humans, even during "cold" periods, as the mountains represented a refuge for life. The caves protected humans, and even once they no longer dwelled there, preserved human DNA for tens of millennia. At this point, you may be feeling a little baffled: Since Denisova and Neanderthal shared room and board and conceived children together, are they really two distinct species? Are we dealing with a paleontological phantom or a genetic chimera? Come to think of it, what even is a species?

The European Neanderthals were strikingly muscular
and robust. Their cousins, the Denisovans, likely
resembled them in this respect.

3

Denisova: A Human Species?

The inventor of the scientific classification of species was a Swedish man called Carl Linnaeus (1707–1778). Ironically, Linnaeus himself began life without a proper name of his own; in the eighteenth century, most Swedish families did not have a surname. Thus, Linnaeus's father was simply known by the name of Nils Ingemarsson, or "Nils, son of Ingemar." Linnaeus in turn, began his life as Karl Nilsson, or "Karl, son of Nils." It was only later that he adopted the name by which we now know him, in compliance with university regulations requiring students to bear a Latin surname. Taking inspiration from the name of his family farm, Linnegård, he became known to the world as Carl Linnaeus.

By this point in the book, you've already come across Linnaeus's taxonomy many times. The full name of our species is *Homo sapiens* (i.e., "wise man," an expression coined by Linnaeus in 1758); that of our Neanderthal siblings is *Homo neanderthalensis*. In scholarly terms, Linnaeus's classification (there were others, as we shall see) is referred to as "binomial nomenclature," because it combines two names: that of the genus, written in Latin with a capital letter, and that of the species, written in Latin in lowercase. The use of Latin requires that the name be indicated in italics.

When a species is named for the first time, it is customary to add the name of its founder and the year it was named to the pair of Latin terms (e.g., *Homo neanderthalensis*, King 1864). The Linnaean classification solves a major problem: It provides all the world's scientists with standardized species names, which is essential, since in each of the many human cultures, animals, and plants have different names.

As for Denisova, its binomial could be either be *Homo denisovensis* or *Homo altaiensis*—i.e., "Altai Man." To date, the International Commission on Zoological Nomenclature has not named the species, a silence that reflects the confusion surrounding the status of this human form. Denisova materialized from the little finger of an adolescent girl for whom a body has not yet been found amongst the other fossils in the cave. For members of the commission, this is quite the conundrum.

For us too: From the first pages of this book, we've been telling you about a new human form that everyone agrees exists, but that nobody wants to officially recognize as such. This tug-of-war leaves us in much need of clarification. What do geneticists really mean when they describe Denisova as a "new type of human," if everyone understands them to mean a new species? This seemingly banal question leads us to interrogate the notion of species, its origins, and its limitations. The concept of species existed before paleogenetics, at a time not so long ago when the only way of identifying them was through zoological or paleontological observations. Does a species that has been identified "solely on the basis of genes" merit its place in the Linnean taxonomy? There is no clear answer.

Portrait of Carl Linnaeus (1707–1778), published in the
first edition of *Philosophia Botanica* (1751)

A Catch-All Term

What is a species? The word is used in a variety of ways in everyday language. Depending on the context, it can either refer to a "kind," a "variety," or a "type," with no precise meaning. This has not been the case in the scientific community for a long time. The notion of species first appeared in 1686, in botanist John Ray's *History of Plants*. Ray used the Latin term "species," which then became integrated into the English language: "No surer criterion for determining species has occurred to me than the distinctive characteristics, which are perpetuated in propagation from seed."

The word "species" therefore emerged from the delicate science of botany, in which plants are distinguished on the basis of very precise, sometimes extremely subtle differences in characteristics. For seventeenth-century botanists, a species is, in short, a group of plants characterized by certain traits, which are transmitted together as a whole. This definition was soon extended to animals, which then raised the prickly question of how to classify all types of living organisms.

All human cultures have developed systems for classifying life forms. The principles of these systems differ, and they are often mutually incompatible. In the fourth century BCE, in his *History of Animals*, Aristotle made the distinction between "sanguine animals" (vertebrates) and "bloodless animals" (invertebrates). This distinction may come as a surprise, since we now know that invertebrates have blood (often blue) making them no less noble forms of life than others.

After him, in the eighth century, the Arab theologian and naturalist Al-Jahiz (776–868 CE) attempted to establish a classification of animals based on morphological, physiological, and ethological (behavioral, such as locomotion or feeding) criteria, which he presented in his *Kitāb al-hayawān* ("Book of Animals"). In it, he described and analyzed various major animal species, but without classifying them in any truly systematic way. Instead of species, he defined "units" according to certain criteria such as shape and reproductive isolation—i.e., the fact that individuals of a given group can only reproduce with other members of that group. Only later would it become apparent just how relevant this theory was.

Fourteenth-century copy of a panel from Al-Jahiz's manuscript, from the Biblioteca Ambrosiana in Milan

The modern classification of living organisms is based on Linnaeus's binomial nomenclature, which classifies species by kinship. The system is made up of interlocking categories like a set of Russian dolls. Species is the base unit. It belongs to the parent category of genus (which groups together "close relatives"—i.e., those similar in form) which in turn fits into the category of family ("distant relatives," or ancestors that share a family resemblance). The next two ranks in the hierarchy are order (forms that share anatomical characteristics), which in turn belongs to class (forms that share an anatomical plane).

Where do modern humans fit into this grand classification? We belong to the class "Mammalia," the order "Primates," the family "Hominidae" (hominids), the genus "*Homo*" and the species "sapiens." What about *Homo neanderthalensis*? They belong to the genus *Homo*, which make them our close relative. Should Denisova be confirmed to be a species, it would also be classified under the *Homo* genus.

A Tale of Two Lions

Linnaeus and his successors relied essentially on morphological criteria to classify organisms into these different categories. But isn't there a more objective biological criterion than form? In science, everything comes down to consensus—i.e., the collective acceptance of a proposition, which might only be the best one temporarily, until new information leads us to revise this judgment. The most relevant biological definition of species today was created by German–American biologist Ernst Mayr (1904–2005).

British physical chemist Rosalind Franklin (1920–1958) at the microscope: It was her X-ray diffraction images that revealed the "double helix" structure of DNA, which was later deduced by James Dewey Watson and Francis Crick (1916–2004), earning them the Nobel Prize in 1958. To this day, Rosalind Franklin remains forgotten whenever the discovery of DNA is mentioned, despite having played a crucial role.

Mayr began his career as an ornithologist in New Guinea and the Solomon Islands. The work he carried out there on the genetic drift of founder populations and the phenomenon of the emergence of new species (known as speciation) led to his appointment in 1931 by the American Museum of Natural History in New York as curator of the ornithological collections. In this position, he was able to continue his work on evolution and the classification of species. During the 1930s, some of the greatest biologists, naturalists, and paleontologists fleeing Nazism took refuge at the American Museum of Natural History. Alongside Mayr, they created a space for reflection on evolution, systematics, and the notion of species.

In this intellectual space, Mayr, together with British biologist Julian Huxley (1887–1975) and American paleontologist George Gaylord Simpson (1902–1984), developed the most widely accepted neo-Darwinian theory of the twentieth century. The greatest biological minds of the time worked on this great synthesis of ideas, taking Darwin's arguments and further developing them. Their theory is incredibly spartan, since it is based on two evolutionary mechanisms alone: natural selection and the random occurrence of mutations in genetic material. While natural selection seemed an obvious criterion in the 1950s, taking mutations as the second was pioneering given that the structure of DNA had only just been published. This conceptual simplicity explains the name given to this neo-Darwinian theory: "Modern Synthesis."

Using these two principles alone, modern synthesis is able to explain the origin and evolution of billions of traits in living beings, and provide descriptions in fields as diverse as genetics, zoological classification, paleontology, comparative anatomy, and so on.

It was in the context of the development of this extremely significant theoretical framework that Ernst Mayr, in his famous book *Systematics and the Origin of Species*, defined a biological species as follows in 1942: "A population or set of populations whose individuals can actually or potentially reproduce with each other and produce viable and fertile offspring, under natural conditions."

Formulated thus, the concept appears to be simple. But its interpretation is anything but. For starters, a species is a large group of living forms that are able to pass on genes through reproduction. A biological species is first and foremost a given reproductive community, whose members are reproductively isolated from other reproductive communities.

According to Mayr's definition, it is essential for the individual born of this act of reproduction to be viable and fertile "under natural conditions." This point may seem like a mere detail, but it isn't; in fact it's the reason why the horse and the donkey are two different species. Their hybrids—the mule (female horse + male donkey) and the hinny (male horse + female donkey)—are sterile. It is therefore the so-called interspecies barrier—the fact that certain individuals cannot reproduce with each other—that distinguishes species from one another. In the case of the donkey and the horse, the barrier is not to the possibility of reproduction, but rather to the fecundity of the offspring. It is this form of reproductive isolation, separated by one generation, which distinguishes the gene pools of the donkey (*Equus asinus*) and the horse (*Equus caballus*).

Speciation in Question

Mayr's definition has an essential corollary: It explains the birth of a new species, or speciation. The requirement that the offspring of two individuals must be fertile in order for them to be considered the same species is not always easy to apply. Even if individual F1 is able to reproduce fertile offspring with F2, and F2 with F3, etc., up to FN, it is not necessarily the case that FN will be able to produce fertile offspring with F1. This phenomenon results from the existence of "inter-species barriers," which can be geographical (mountains, seas, etc.), biological (specific feeding habits, behavior, etc.), or even cultural (in the case of social animals).

Thus, a group isolated from the rest of the population by unsurmountable geographical barriers—a great distance, a sea, a river, a mountain range, etc.—is likely to evolve into a new species. This phenomenon has been studied, for example, by the Austrian ethologist Konrad Lorenz (1903–1989) in the case of seabirds of the same origin separated by distance.

But even speciation is a complex phenomenon, since different populations have been observed to remain interfertile for a considerable time. A well-known example is the *Panthera leo persica*, the Asian lion, which is able to reproduce with its African cousin *Panthera leo leo*, but has distinct traits, such as a thicker tuft at the end of the tail, a smaller mane, a different way of life, and so on. Such cases are defined as subspecies, which explains the nuances *persica* or *leo* added to the species name.

Species in Paleontology

Let's return to our fossils. Given that the concept of species is based on fertility, it is impossible to apply it to extinct species. How were paleontological species defined before the emergence of paleogenetics? Paleontology (the study of what has existed before the Holocene) and in particular, paleoanthropology (human paleontology), makes a distinction between "morphospecies." This taxonomy creates groups based on similarities and differences in features. All individuals with the same "specialist" set of structural (number of limbs, fingers, etc.) or anatomical (shapes of bones, teeth, etc.) traits are placed within the same morphospecies.

The advantage of this mode of classification is that it can be applied systematically. Given the millions of fossils in existence, this has an enormous advantage. The disadvantage lies in the subjectivity that governs the choice of criteria, as illustrated by the story of the discovery of Neanderthal. In 1856, the first people to see the fossil remains in the Neandertal valley successively identified a cave bear (discovered by excavation workers), a wounded Cossack who had come to die in the cave (discovered by the first scientist) and a *Homo sapiens* suffering from rickets (discovered by the second scientist).

It was in Germany, in this quarry near Mettmann in the Neandertal valley, that the Neanderthal holotype—the first recognized fossil of *Homo neanderthalensis*—was discovered in 1856. This famous fossil is now preserved in the Rheinisches Landesmuseum Bonn.

In France, Marcellin Boule, professor of paleontology at the Muséum National d'Histoire Naturelle, was the first to carry out a systematic study of a complete Neanderthal skeleton, following the 1908 discovery of "The Old Man," in La-Chapelle-aux-Saints in Corrèze. *L'homme fossile de La Chapelle-aux-Saints*, the anatomical description published by Boule between 1911 and 1913 in *Annales de paléontologie*, is considered so outstanding that it remains a benchmark to this day. That being said, his interpretations were influenced by the prejudices of his time, many of which are shocking today: "The first striking thing about the head is its considerable dimensions, particularly in relation to the small stature of its former owner. It is also striking for its bestial appearance, or to put it more accurately, the combination of its simian or pithecoid [ape-like] features."

Bestial? Neanderthal? This is certainly the reputation Neanderthals have since garnered among the general public, but

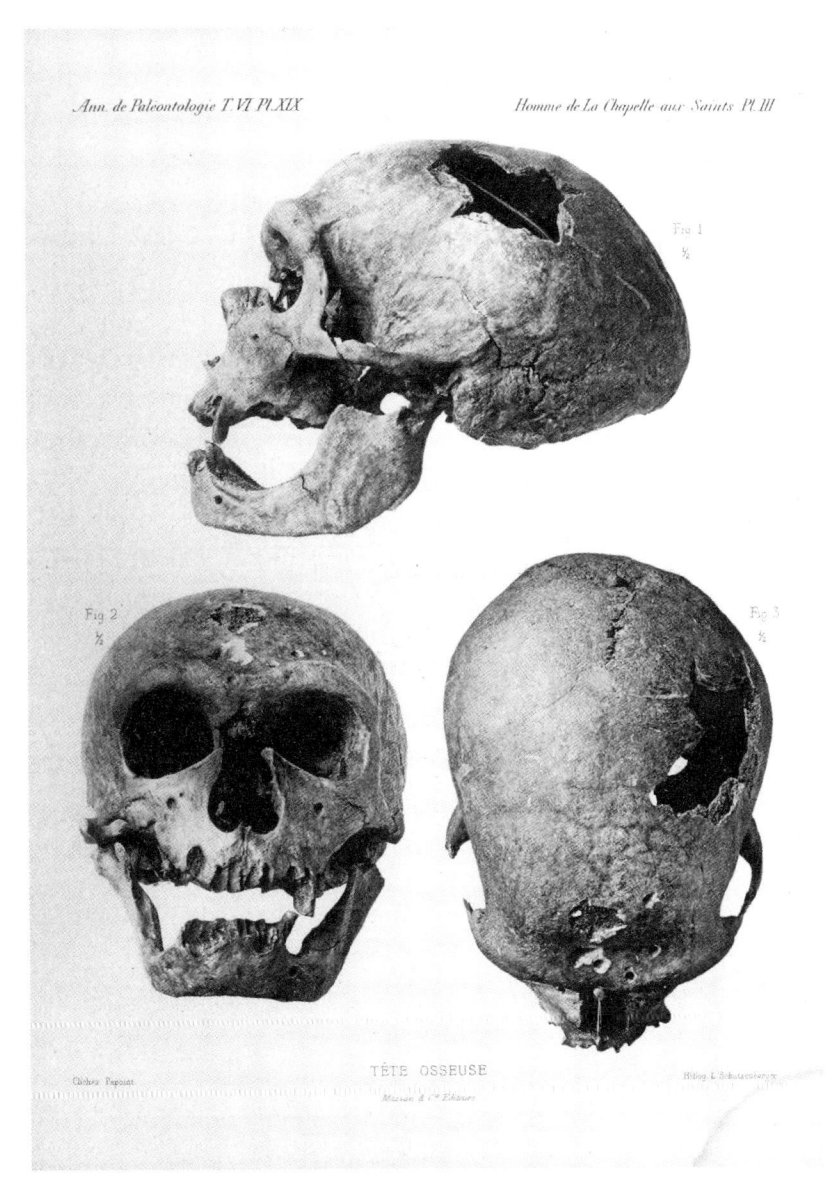

Ann. de Paléontologie T. VI Pl. XIX

Homme de La Chapelle aux Saints Pl. III

The skull of the Neanderthal known as "The Old Man," as presented by Marcellin Boule in *Annales de paléontologie* between 1911 and 1913

not in the field of prehistory, since we have come to realize that they appeared, evolved and lived for almost as long as their contemporaries, *Homo sapiens*.

What's more, their material culture—all the objects and other materials that provide evidence of their behavior—is no more primitive than that of *Homo sapiens* of the same stage of evolution. In fact, around one hundred thousand years ago in the Levant, their material cultures were equal. The Neanderthals' "bestial" traits are largely explained by the cold environment in which they evolved in Europe.

It would take too long to detail here the many features that separate Neanderthal from Sapiens, so let's content ourselves with a brief presentation of the head, because, as we shall later discover, the identification of the characteristics of the Denisovan skull played a crucial role in characterizing the Denisovan morphospecies. But let's not get ahead of ourselves. What was the reason behind Marcellin Boule's description of the Neanderthal skull as "bestial"? Firstly, Neanderthal skulls are mainly "rugby ball"–shaped (i.e., long), while Sapiens skulls are "soccer ball"–shaped (i.e., round). Neanderthal cranial bones are low and elongated from front to back, while Sapiens develop high cranial bones. Secondly, the Neanderthal face is "snout-like" (i.e., at the front of the skull), whereas the Sapiens face is at the base of the forehead, below the skull.

Perhaps you get a sense of strangeness from these few Neanderthal traits. No doubt you'd agree that *Homo neanderthalensis* is a separate morphospecies from *Homo sapiens* from its skull alone. But this is only subjective. The strangeness these differences evoke in our minds is perhaps no greater than the equally real sense that might be felt by an Australian Aboriginal person upon contemplating the narrow noses and pale faces of Europeans, for example.

As the case of Neanderthal illustrates, the choice of which bone features should characterize a particular morphospecies plays a very central role in paleoanthropology. It's so crucial that it remains the subject of long and bitter debates on the origin and evolution of

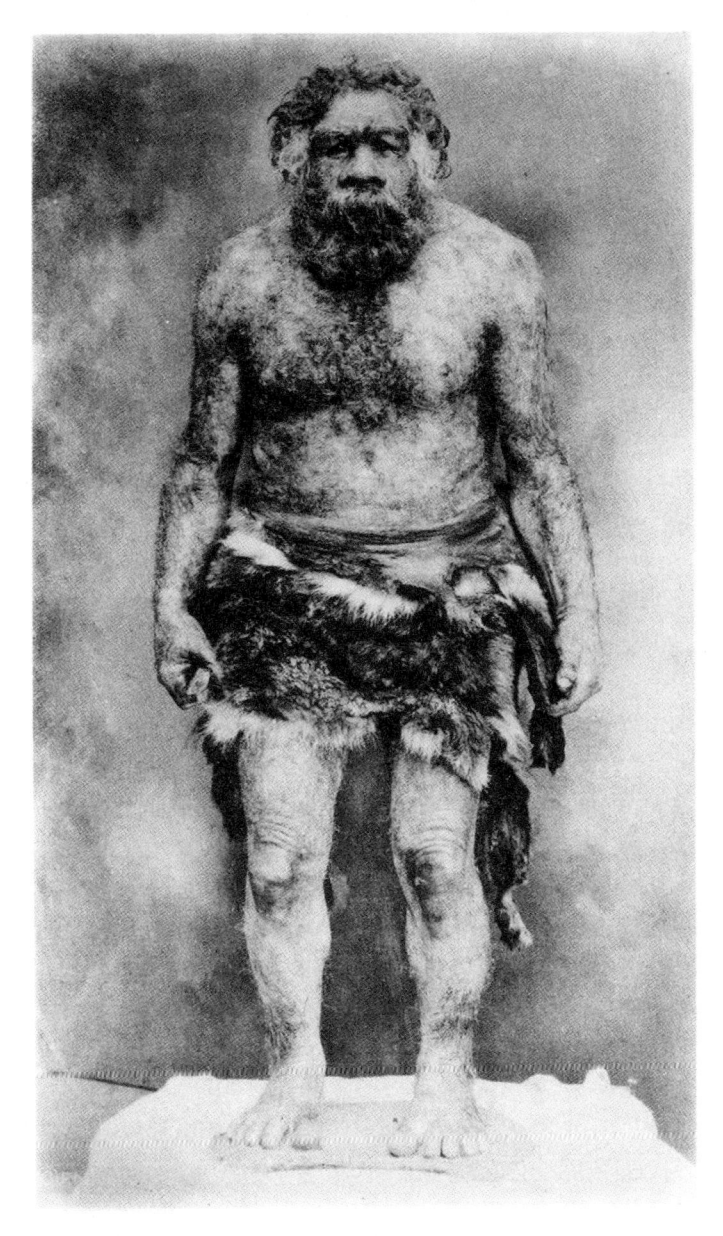

This reconstruction of a "Neanderthal Man" by Frederick Blaschke, as imagined in the 1930s, seventy years after the discovery of the holotype, was exhibited at the Field Museum of Natural History in Chicago.

traits: Are they produced by speciation without adaptive benefits, or are they the result of adaptations to particular conditions? The interpretation of each trait leads to intense clashes among researchers. Having studied Neanderthal bone traits time and again, paleoanthropologists ultimately named them a species in their own right. Paleontologists have reached an impasse with Denisova, because the few teeth and bone fragments available seem to be devoid of any distinguishing traits.

In short, this is an unprecedented moment in paleoanthropology. For the first time in the history of science, a form has been distinguished by its genes and nothing else. Can it therefore be considered a species?

Species According to DNA

In a word, yes: Denisova is a species. But the concept of "genetic species" is highly technical. It posits that individuals in a group belong to the same species if there are significantly fewer genetic variations within this group than there are between this group and all others. It might sound fairly simple, but this is also relative. To get an idea of what this means, let's look at what percentages of DNA are shared between different living forms. Humans, for example, share around 35 percent of our DNA with the *Narcissus jonquilla*, the daffodil, but practically 98.8 percent of DNA with *Pan troglodytes*, the chimpanzee. However, this method of comparing DNA is biased, because it's mainly the genes—the DNA sequences responsible for protein synthesis—that are taken into account. A genome is made up of "protein-coding" DNA and "non-coding" DNA. Contrary to what we might have learned at school, one gene can be involved in several functions, and several genes in the same function. We don't know what roles many of our genes play, which means that we don't yet know exactly what makes two species genetically different.

And yet, all coding DNA put together represents only 1.5 percent of total DNA. Even if this percentage seems low, the comparison of fourteen thousand genes in humans and chimpanzees has revealed five hundred different genes, involved in reproduction, immune defenses, sensory perception (notably the senses of smell and hearing), and of course, body shape.

At the very least, we can say that the apparent characteristics of the chimpanzee, whose DNA is only 1.2 percent different from ours, are enough to distinguish us from them. It's easy to understand why, even if only a tiny fraction of the chimpanzee's DNA differs from ours, this is enough to make us two different species and place us in two different genera.

Where do the genetic differences between chimpanzees and humans come from? They result from the accumulation in our lineages of "mutations"—i.e., heritable modifications of the DNA. Some mutations are "background" (accidental), but many occur under the influence of chemical agents (e.g., pollutants), environmental agents (e.g., heat, ultraviolet rays), and so on. These are known as gene mutations. Most mutations have little effect (these are known as neutral mutations), but those that do are subject to one of two different mechanisms: negative selection, which eliminates harmful mutations, and positive selection, which increases their genetic imprint by multiplying their repetitions in the DNA sequence.

In practice, to demonstrate the effects of mutations, geneticists compare DNA sequences base by base. In doing so, they have been able to determine that an average of seventy mutations occur during the formation of a new nuclear genome. Even if the number of mutations varies from one part of the genome to another, this average value gives the "average mutation rate per generation." This rate plays a very important role in paleoanthropology, as it is one of the cogs in the "molecular clock." After counting the mutations that distinguish two human forms, paleogeneticists divide their number by this rate to obtain the time elapsed since their divergence.

The Genetic Distinction Between Sapiens and Neanderthal

The molecular clock suggests that Sapiens and Neanderthal separated over half a million years ago. But what does genetics say about the differences between Neanderthals and us? Quite a lot, since paleoanthropologists have been accumulating observations on the bone traits of *Homo neanderthalensis* for 170 years. A base-by-base comparison of the DNAs of two of our contemporaries reveals that they differ by only one nucleotide in 1,000. That of one of our contemporaries and that of a Neanderthal by one nucleotide in 1,300. This means there is only a 0.03 percent difference between Sapiens and Neanderthal DNA. To understand what these figures mean, let's recall that Eurasian genomes can contain entire sequences in common with Neanderthals, which represent between 1.8 and 2.6 percent of our DNA.

Both the low level of genetic variability within our species and the modest variability that separates us from Neanderthal and Denisova set us apart from other great apes. For example, the genetic diversity of gorillas is twice as great, and that of chimpanzees and orangutans three times as great. All this illustrates how an anatomical analysis and a genetic analysis have very different meanings. While the Neanderthal form is quite distinct from the Sapiens form in its appearance, it is very close to our own in its genes, at least close enough for the two species to be interfertile.

A Denisovan Species?

With all this in mind, it's not hard to understand the questions the geneticists at Harvard University and MPI-EVA were asking themselves once the *Denisova 3* genome had been sequenced. As in the case of Neanderthal, they found that the differences between Denisovan DNA and that of present-day humans were too numerous for the genomes of Denisova and Sapiens to be considered identical. They also concluded that Neanderthal and Denisova were closer to each other than each of them is to Sapiens (though all three are still very close to one another).

Is Denisova a species or not? The genetic proximity between Neanderthal and Sapiens clearly illustrates that DNA alone is not enough to provide a clear definition of species. *Homo neanderthalensis*'s status as a species is widely accepted primarily because paleoanthropologists have been observing bone differences between Neanderthals and Sapiens for the past 170 years. Once Denisovan fossils have been identified, the same is likely to apply.

Another key factor is that, since *Homo neanderthalensis* and *Homo sapiens* were more interfertile than donkeys and horses are with each other, they do not constitute two biological species in Ernst Mayr's sense, but rather two "biological subspecies."

Establishing genetic differences is important, but they are so few in number that the study of apparent (phenotypic) traits remains an essential factor in deciding whether a human form constitutes a morphospecies. Still, genetic analyses of the kind that led to the revelation that Neanderthal and Sapiens had bred with one another are essential for reconstructing the demographic history of populations and understanding the ways in which they adapted. As we will see, the study of Denisovan genes is no exception: It tells us a great deal about the history of the species.

Did the three human species cross paths? If so, would they have enjoyed getting to know one another? (A Sapiens is woman pictured on the left, a Neanderthal opposite, and a Denisovan on the right.)

4

The Story of Denisova as Told by Genes

Before the publication of *On the Origin of Species* in 1859, Charles Darwin (1809–1882) was a young country gentleman who hunted and observed a great deal of animal life. This is how he came up with his great idea: Nature acts like a breeder, selecting lineages endowed with particular traits. In domesticated species, this might be the running speed of a horse, the meat mass of a beef bull, the number of grains per ear of wheat, etc. In nature, it might be the camouflaging coat of a fox, the swimming speed of a tuna, but also the color of a peacock's feathers.

This is known as "natural selection" and has three consequences, which Darwin characterized as such: All individuals of a sexual species are different from one another; the traits that distinguish them from one another are transmitted in part during reproduction; the individuals within the species are subject to constraints in their environments—known as "selection pressures"—which have demographic effects, because they affect their survival and reproduction. These selection pressures encourage the spread throughout the population of advantageous mutations (the positive selections mentioned before), which are conducive to survival and reproduction.

It Must Be in Their Genes....

This brief outline of Darwin's theory is important because it plays an essential role throughout the rest of this book. It will help us to interpret what we have learned from studying the genes of Sapiens, Neanderthal, and Denisova. Genes can tell us a lot about the origins of fossil populations, their physiological traits, and how they adapted to their environments, including microbes. While it's technically possible to identify these biological traits straight from the Denisovan genome, in practice, determining their influence is no simple matter.

To do this, we'd need to be able to take into account the influence of non-coding sequences (around 98.5 percent of the human genome) and their repeats. We also need to know all the mechanisms controlling genome expression, which is known as epigenetics. This part of our heredity is poorly understood but plays many roles in the differences between Sapiens, Neanderthals, and Denisovans. We don't know how the thousands of genes required to produce the Denisovan phenotype work together.

As we'll see, some phenotypic traits can be deduced from Denisovan DNA, but given the lack of fossils, it's a tricky task to determine these with any certitude. What can the Denisovan genome tell us? For starters, it tells us about its divergence from the Sapiens and Neanderthal genomes. When did this separation occur? On which continent? Taking a close look at the Denisova genome can help us to explore the history of its lineage.

The First Clues

As early as 2010, when Denisova was discovered, the first clues began to emerge. Right away, MPI-EVA paleogeneticists observed that the Denisovan lineage possessed mitochondrial DNA that differed from that of Altai Neanderthals and Sapiens. At the head of their article in *Nature*, they presented this new lineage as follows: "It represents a hitherto unknown type of hominin mtDNA, that shares a common ancestor with the mitochondrial DNA of anatomically modern human and Neanderthal mtDNA about 1.0 million years ago."

To recap, "hominins" is the name that refers to all the forms of the human lineage since its separation from that of the chimpanzee. However, in a later publication, researchers estimated the mitochondrial divergence between the new Eurasian hominin, *Homo sapiens*, and Neanderthals at around 750,000 years ago, around the same time as the human *Homo rhodesiensis/heidelbergensis*'s departure from Africa. The fact that the

Portrait of the naturalist Charles Darwin (1809–1882) by Julia Margaret Cameron: Darwin's book *On the Origin of Species*, published in 1859, gave rise to the theory of the evolution of all living beings.

mitochondrial divergence dates back to this time would suggest that the new hominin had inherited its DNA from the same *Homo rhodesiensis/heidelbergensis*, the common ancestor of Sapiens and Neanderthal.

This was confirmed in 2013 by the remarkable sequencing of mitochondrial DNA miraculously preserved for 430,000 years at the bottom of the Sima de los Huesos pit cave in Atapuerca, Spain—the oldest known human mitochondrial DNA. This discovery shed light on the MPI-EVA's findings on mitochondrial divergence. The fossils excavated from the pit were in fact *Homo heidelbergensis* in the process of Neanderthalization (that is, becoming Neanderthal). They were so well preserved in the depths of Sima de los Huesos that, in an unprecedented feat, Svante Pääbo and his team succeeded in extracting 98 percent of the mitochondrial DNA from a femur . . . and this DNA turned out to be the same type as that of *Denisova 3* (a Denisovan). The fact that the mitochondrial DNA discovered at Sima de los Huesos and Denisova was so similar suggested that it was that of the common ancestor of Neanderthals and Denisovans: *Homo heidelbergensis*.

At the time, it was hard to believe: Why did pre-Neanderthals in Spain, who were in the process of Neanderthalization, not have the same mitochondrial DNA as subsequent European Neanderthals? Where did the mitochondrial DNA of the more recent Neanderthals come from? The only possible explanation was that Neanderthals had mixed with another human more recently than 430,000 years ago, receiving a new mitochondrial DNA, different from that of Sima de los Huesos and *Denisova 3*. This hypothesis was confirmed in 2017 when geneticist Cosimo Posth from the Max Planck Institute for Human History in Tübingen demonstrated that the human mitochondrial DNA found in a Neanderthal tibia unearthed in the Hohenstein-Stadel cave in Swabian Jura, Germany, was the same as that of archaic *Homo sapiens*, which Neanderthals undoubtedly frequented very early on in the Levant. The mitochondrial divergence evidenced by *Denisova 3* and Atapuerca tallies with evidence (the discovery of a hand axe) that a human wave appeared in Europe after seven hundred thousand years ago. As its members evolved into Neanderthals, hand axes became increasingly sophisticated. The only tool found at Sima de los Huesos was a magnificent pink hand axe—nicknamed Excalibur—which shows that the *Homo heidelbergensis* from Sima de los Huesos belonged to this wave. It is therefore clear that when this wave arrived in Europe, it was carrying *Homo heidelbergensis* mitochondrial DNA, which was later substituted by archaic Sapiens DNA.

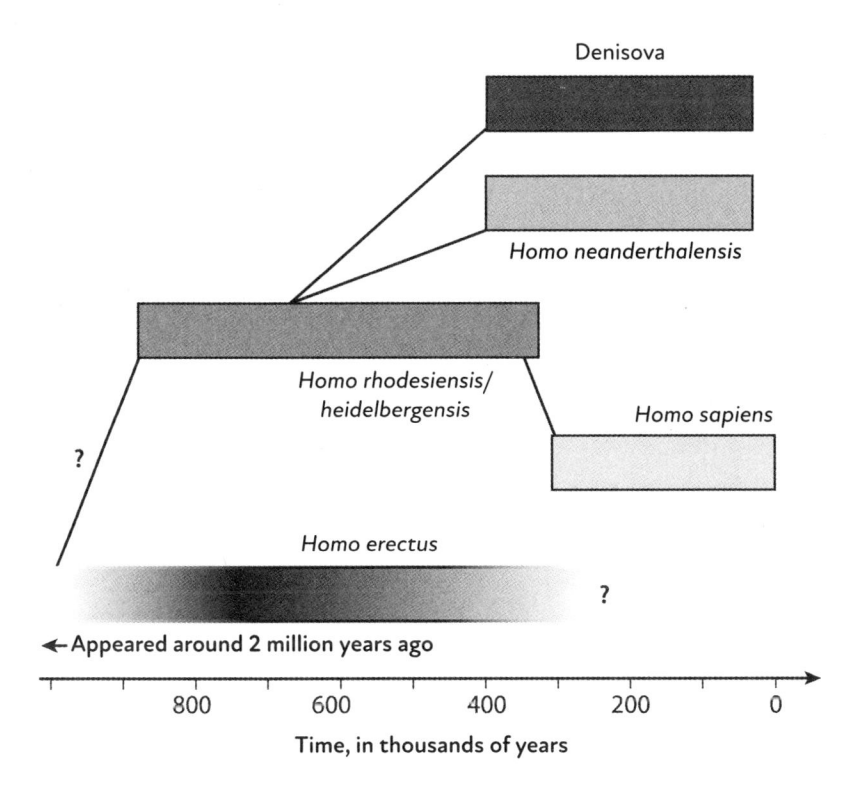

Summary of human evolution in Europe, Africa, and Asia over the last 800,000 years

And what about the nuclear DNA contained in cell nuclei? As early as 2010, David Reich's team at Harvard University succeeded in making a first, fairly rough reading of the nuclear DNA sequence of *Denisova 3*. When this was compared with the Neanderthal nuclear DNA available at the time, it became clear that the divergence between the Neanderthal and Denisovan lineages was only around 400,000 years old. This was confirmed in 2014 on the basis of a new, much higher-quality sequencing of *Denisova 3*, which had been carried out in 2012. By comparing the resulting nuclear genome with that of Altai Neanderthals (*Denisova 5*), MPI-EVA researchers calculated that Denisovans and Neanderthals diverged between 473,000 and 381,000 years ago.

Denisova's Siblings

What does all this tell us? That Sapiens, Neanderthal, and Denisova evolved from *Homo heidelbergensis*. In Eurasia, the Neanderthal and Denisovan lineages diverged earlier than four hundred thousand years ago, then evolved separately over the same periods: One became Neanderthal, in western Eurasia, and the other Denisovan, in eastern Eurasia. This is a hugely significant discovery. Neanderthals had an eastern sibling that looked just like them. This observation plays such a crucial role in solving the Denisova enigma that it bears repeating.

It's customary to refer to the evolutionary process that produced Neanderthal as "Neanderthalization." We now believe that, while Neanderthalization was happening in Europe, a comparable evolutionary process was taking place in Asia that we refer to using the neologism "Denisovation." The fact that conditions in eastern Eurasia are, and always have been, comparable to those of western Eurasia (aside from a few notable differences), is necessarily reflected in the convergences and divergences between Neanderthalization and Denisovation.

A Question of Depth

This is not the only thing we learned from the much higher quality sequencing of *Denisova 3* carried out in 2012. The research carried out in 2010 had been at a low sequencing depth of

1.9x. You might be wondering what is meant by depth—a very important term. Mitochondrial DNA represents 0.0005 percent of the total number of nucleic bases in the human genome, while nuclear DNA accounts for 99.9995 percent. The nuclear genome therefore contains far more information than the mitochondrial genome—almost all, in fact. In bone samples, there is an enormous amount of foreign DNA (exogenous DNA) that comes with the DNA to be reconstituted for research purposes (endogenous DNA). Exogenous DNA comes from the numerous necrophagous organisms that attack the remains after death. These include bacteria, soil fungi, etc. The fossils can also be contaminated by a large amount of human DNA when carelessly handled.

A depth of 1.9x means that, when the sequence libraries are built, each piece of endogenous or exogenous DNA is read 1.9 times on average. This is a very low number. When a reconstitution of the complete sequence is carried out with the help of bioinformatics, this lack of information results in statistical uncertainties, prompting researchers to attempt to gain a more reliable result. This proved to be the case in 2012, when Matthias Meyer from MPI-EVA and his team presented the results of a very deep sequencing (30x) carried out on 40 milligrams of the *Denisova 3* phalanx. The team made spectacular progress by developing a method that increases the richness of the sequence libraries, making it possible to read each base at least thirty times—a method that has since become widely used.

An Unknown Human Form

The resolution and reliability achieved with this depth have yielded several striking discoveries. The first is that Matthias Meyer and his team were able to ascertain that Denisovans and Neanderthals were of African origin. In support of this finding, Silvana and the rest of her team at Aix-Marseille University used the *Denisova 3* genome, which is now known almost in its entirety, to analyze seven blood group systems, including two that are commonly used in blood transfusions today: ABO and Rhesus. The list of blood group systems prevalent in present-day Africans suggests that the blood groups of Denisova and Neanderthal already contained all the combinations known in the cradle of humanity today. The inevitable conclusion is that they both have African origins. This is a well-known fact in the field of prehistory, but gaining further confirmation through blood analysis is important, given that some paleoanthropologists hypothesized that Denisova was of exclusively Asian origin.

Secondly, it appears that Denisova also interbred with another Asian form: The genome contained in *Denisova 3*'s little finger

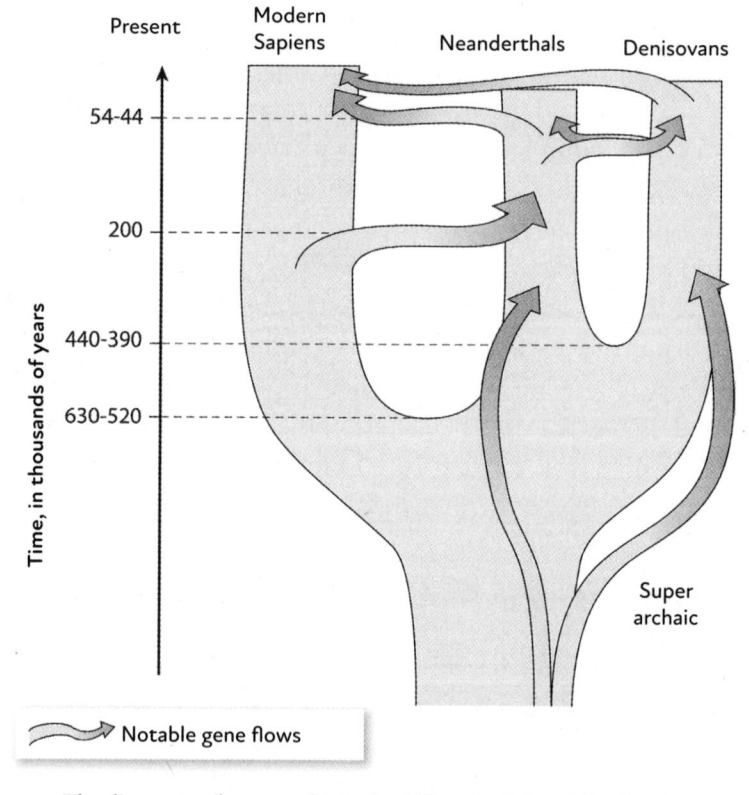

The divergences between the Sapiens, Neanderthal, and Denisovan branches, as revealed by the study of mitochondrial and nuclear DNA (particularly the Y chromosome)

phalanx in fact carries some DNA sequences inherited from an unknown, so-called super-archaic hominin. According to the molecular clock—i.e., the number of mutations present in the four million or so bases making up these sequences—the ancestral lineage of Neanderthals, Denisovans, and Sapiens diverged from that of the "super-archaics" between one and one and a half million years ago. Who could this unknown hominin be? *Homo erectus*, no doubt, based on our current state of knowledge, but we can't be sure.

The third striking discovery is the low heterozygosity of Denisova 3. The quality of the genome obtained for this individual is high enough to make it possible to estimate the proportion of genes whose paternal and maternal versions (the alleles) differ—i.e., its "heterozygosity." Within a population, greater heterozygosity means the circulation of more versions of each gene—i.e., greater biodiversity, which is conducive to adaptation. Populations with low genetic diversity therefore beget individuals with low heterozygosity, as is the case for Denisova 3.

The final conclusion is that it appeared some Melanesians (Papuans) had amassed between 4 and 6 percent of Denisovan DNA, which is enormous. In 2010, this result, based on the sequencing of just five *Homo sapiens*, stunned the scientific community, especially since among these individuals, Han Chinese genomes (those of the majority Chinese ethnic group) seemed to have only traces of Denisovan DNA, far less than the Papuans. This finding threw researchers off, given how far Melanesia is from Denisova's cave . . . but that's another enigma for another chapter. Suffice it to say that at the time, the significant level of interbreeding between Denisovans and Sapiens so far from the Altai led to hopes of finding Denisovan fossils and genomes outside of the Denisova Cave.

Denisovans: A Boneless Lot

Sadly, luck was not on the paleoanthropologists' side. For almost ten years after the 2010 discovery, the Denisovans remained boneless, apart from a few fragments less than a centimeter in size, a rather uninteresting specimen of a cranial vertex, and a single tooth. That was until 2019, when prehistorian Zhang Dongju from the Lanzhou University, together with the paleoanthropologist Jean-Jacques Hublin, then at MPI-EVA, launched an investigation to prove that a mandible discovered in 1980 in the Baishiya Karst Cave, in the far northeast region of the Tibetan plateau, was Denisovan. Zhang Dongju assembled a team to try and extract Denisovan DNA from the cave sediments and their mission was a success.

The Chinese team used a method at the Baishiya Karst site that had been developed at Denisova cave by Viviane Slon of MPI-EVA alongside other scientists. In 2017, the researchers had collected 85 sediment samples from caves inhabited by Neanderthals or Denisovans, from which they subsequently attempted to extract mitochondrial DNA. Surprisingly, while 1 milligram of ancient bone can usually be expected to contain between 34 and 9,000 fragments of mitochondrial DNA, 1 milligram of sediment here contained between 30 and 4,500 fragments of mitochondrial DNA belonging to a whole range of species from mammals, through fungi, to plants. Of course, only the minority of this was human mitochondrial DNA, but Viviane Slon's paleogeneticists were still able to identify it in the sequence libraries obtained.

Zhang Dongju's team did the same: After extracting mitochondrial DNA fragments preserved in sediment samples from the Baishiya Karst Cave, they sequenced them to obtain sequence libraries. Once the human and non-human mitochondrial DNA had been separated, the researchers compared the preserved sequences with those of Sapiens', Neanderthals', and Denisovans' mitochondrial genomes, and identified Denisovan mitochondrial DNA at various depths in the stratigraphic pile. It emerged that Denisovans had stayed in the Baishiya Karst Cave from time to time between 160,000 and 45,000 years ago. In short, another cave of Denisovans had been discovered.

It would appear that certain Denisovans had lived at high altitudes, as confirmed by the fact that *Denisova 3* possessed a version of a gene—the EPAS1 allele—that would have prevented them from suffering long-term altitude sickness and its attendant headaches, fatigue, nausea, shortness of breath, confusion. . . . EPAS1 is involved in the body's physiological response to oxygen concentration, which not only improves oxygen circulation in the body, but also plays a role in heart development and function. This allele is therefore very useful for life in the Tibetan plateau, which at an average altitude of 14,764 feet (4,500 m), can cause low oxygen levels. This version of the EPAS1 gene was discovered in *Denisova 3*, though the location of the Denisova Cave (just 2,297 feet/700 m above sea level) does not necessarily require any adaptation to altitude.

If the ancestors of Denisovans lived in the mountains long enough to adapt to life at high altitudes, then the Denisovans must have been present on the Tibetan plateau for several hundred thousand years, as attested by the occupation dates established by Zhang Dongju's team. When the first Sapiens arrived, the Denisovans probably passed on to them the "anti–mountain sickness gene" that today's Tibetans have inherited. If this is the case, there must have been Denisovans living on the Tibetan plateau at the same time as they were living in the Altai. As this vast area covers both temperate China to the north and tropical China to the south, this observation suggests that they also occupied a much larger territory. Suddenly, Denisova's genetic legacy to modern populations made it clear that the species covered a geographical area considerably greater than previously thought. And this was only the beginning.

In order to survive, prehistoric people had to move constantly
depending on the seasons and animal migrations.

5

A Vast Empire in the East

During the Spanish colonization of the Philippines, an indomitable group of natives from the island of Luzon resisted all attempts at sedentarization: the Ayta. According to observational accounts from the time, such as those of Miguel López de Legazpi, the first governor of the Philippines, the "Negritos" (a name originally given to Indigenous peoples by the Spanish) possessed iron tools. The precision and speed with which they were able to use their bows and arrows made them formidable adversaries. During the conquest, they retreated to the wilds of the large mountainous and volcanic island of Luzon, where they continue to live today, virtually cut off from the outside world.

This genetic isolate is a goldmine for geneticists. In 2021, a team of researchers led by Maximilian Larena of Uppsala University teamed up with the Philippine National Commission for Culture and the Arts, local cultural communities, and other actors defending Indigenous rights, to launch a vast study into the genealogy of the country's ethnic groups, particularly the Negrito populations. The findings were incredible: The Ayta Magbukon, a Negrito ethnic group who inhabit the Bataan peninsula on the island of

Luzon, possess 5 percent Denisovan DNA. That's five times higher than the Tibetan population. But how did Denisovan genes end up so far from the Altai cave where they were discovered? Did the Denisovans occupy a much wider and more diverse territory than their adaptation to the cold regions of the Altai and the Tibetan plateau would suggest?

The Ayta Magbukon have an average of 5 percent Denisovan DNA, a legacy of interbreeding between Denisovans and early Sapiens in Southeast Asia.

Before we follow this line of inquiry, let's recap what we already know: Around 750,000 years ago, an ancestral human form (a priori *Homo heidelbergensis*) emerged from Africa and spread across Eurasia. Four hundred thousand years later, two groups had emerged: Neanderthals in the west and Denisovans in the east. We also know that two hundred thousand years ago, Denisovans were probably already thriving in the Altai mountain range and on the Tibetan plateau. As early as 2010, one of the co-discoverers of Denisova, David Reich, guessed that Denisovans might have occupied a much larger territory. Based on his analysis of a limited number of genomes belonging to modern populations, the paleogeneticist had already been able to hypothesize that Denisova had lived as far away as Melanesia, the Southeast Asian archipelago that includes New Guinea. Strangely enough, the percentage of Denisovan genes increased toward the south of the continent. Not only had Denisovans lived in the tropics, they seemed to have firmly established themselves there. However, these impressions were based on the sequencing of five Sapiens individuals alone.

These have since been confirmed. One after the other, several studies have supported the idea of a massive Denisovan heritage in southern Asia, along the entire coastal fringe stretching from India to Melanesia. In West Bengal, India, researchers from the National Institute of Biomedical Genomics in Kalyani—assisted by colleagues from Spain, the UK, the Netherlands, and China—analyzed over 1,739 human genomes from the Indian subcontinent. Using this data, they were able to retrace the population history of the region. Their findings showed that descendants of the first *Homo sapiens* to arrive in the subcontinent mixed with Denisovans already present (as was the case in Southeast Asia). What was particularly interesting about their findings is the fact that Adivasi (i.e., Indian Aboriginal people) possess considerably more Denisovan DNA than Indo-European Indians (numerous in the north), whose Indo-European ancestors arrived in India from the northwest at the end of the Bronze Age. Further to the east of the continent, Mayukh Mondal and his team at the Institute of Evolutionary Biology in Barcelona sequenced the genomes of sixty islanders from the Andaman Islands off the coast of Myanmar. These researchers concluded that these descendants of members of the first Sapiens wave in southern Asia possessed between 2 and 3 percent Denisovan DNA, roughly the same share as that of Neanderthal DNA in European genomes.

Cut from the Same Cloth

Finally, in 2019, a large team of researchers led by Guy Jacobs of the University of Singapore published results on the diversity of

Denisovan genes present in the ancestors of the population that David Reich's team had found have the highest percentage of Denisovan DNA: the Papuans of New Guinea mentioned previously. Their DNA profiles were confirmed, indicating the ancestral presence of Denisovans in the region. But the team made another, even more interesting discovery.

Having selected a set of "blocks" from Papuan DNA that were almost certainly Denisovan, the researchers compared this material with the equivalent data on Denisovans from the Altai. Their analysis revealed that during their expansion across Southeast Asia (starting some seventy thousand years ago) ancient *Homo sapiens* came into contact with not one, but two distinct Denisovan groups. This great discovery suggested that several Denisovan populations had existed in Southeast Asia.

This is not all that surprising, when you think about it. Denisovans evolved for the most part on a continent that no longer exists. For probably more than half of the Pleistocene (from 2.58 million to 11,700 years ago), a huge part of this region of the world—now submerged—was above water. A study by Harold Voris of the Field Museum of Natural History in Chicago shows that, for 40 percent of the last 250,000 years of the Pleistocene, a huge part of the Sunda continental shelf (the extension of the continent below the surface of the Sunda Islands of Sumatra, Java, and Borneo—see page 105) was only 131 feet (40 m) below the ocean's surface. However, during glacial periods, the sea level fell well below this.

During Denisovation and Neanderthalization, the peoples of Southeast Asia lived on a vast expanse of land that had emerged above sea level. This Sunda "continent" extended as far as the Wallace Line, a biogeographical boundary separating the Asian ecozone from its Australian counterpart. During the Glacial Maximums (or coldest periods), the Wallace Line was a deepwater channel running between the islands of Bali and Lombok, to the west of the Philippines. The Wallace Line marks the southern boundary of a large tropical continent, representing the main arrival zone for groups escaping from Africa, since these tropical humans tended to stay in climates to which their bodies were adapted.

The human wave that gave rise to Denisova in eastern Eurasia and Neanderthal in western Eurasia therefore also came to populate the immense Sunda continent, creating a substantial population there. During warm periods, the Sunda continent fragmented into archipelagos. Island populations diverged genetically from one another, which explains why percentages of Denisovan DNA vary from one large island to another in Southeast Asia.

Guy Jacobs's team concluded that the last hybridizations between Sapiens and Denisovans took place between thirty thousand and fifteen thousand years ago, well after the presence of the last Denisovans in the Altai. What's more, according to researchers, the conservation of such high proportions of Denisovan genetic material in Papuan genomes can be explained by its usefulness in pathogen-laden equatorial rainforests, where the specific immune adaptations of Southeast Asian Denisovans must have proved highly advantageous.

Collège de France to the Rescue

This phenomenon is even clearer in the genetic history of the Oceanians, reconstructed over a 10-year period by a large team led by geneticist Lluís Quintana-Murci from the Collège de France and Étienne Patin from the Institut Pasteur in Paris. The researchers sequenced 317 genomes from 20 Pacific populations, from Taiwan to Vanuatu, via the Bismarck Archipelago and the Solomon Islands. Their connection with the Denisovans might seem puzzling, but it becomes clear if we note that these populations were formed through a series of processes that led to the creation of distinct ethnic groups.

This succession of ethnogeneses began in mainland China over four thousand years ago before continuing in Taiwan, then the Philippines around 2500 BCE, before moving on to Indonesia. It then extended over an immense area, where the direct descendants of the first Sapiens to arrive in the region lived (the Negritos and Papuans). It would appear that the ancestors of the Oceanians dispersed in stages throughout Southeast Asia then into the vast territory of Oceania.

What do their genes tell us? Lluís Quintana-Murci highlighted one essential aspect: "We were surprised to find that . . . the heritage of Denisovans varies considerably from one population to another: From practically 0 percent in Taiwan and the Philippines [authors' note: excluding hunter-gatherers], the number rises to 3.2 percent in Papua New Guinea and Vanuatu."

Some forty-six thousand years ago, while still in Asia, the Sapiens ancestors of the inhabitants of nearby Oceania (Papua New Guinea, the Bismarck Archipelago, and the Solomon Islands) and of the Aytas in the Philippines received their first gene flow from a Denisovan population that had separated from the Altai Denisovans over two hundred thousand years ago. The ancestors of the Aytas in the Philippines also received a second flow from another, probably local, Denisovan group. Then, in eastern Asia, around twenty-one thousand years before the present, an ancestral population of modern-day Oceanians received a third gene flow from a group of continental Denisovans before migrating to Southeast Asia. Finally, a last Papuan-specific gene flow is thought to have originated, some twenty-five thousand years ago, from a local Denisovan population separated from Altai Denisovans over four hundred thousand years ago.

In short, four different waves of Denisovans contributed to Oceanian sapiens' DNA. This shows that the numerous Sapiens populations that moved from continental Asia to Oceania brought Denisovan genes from the north (separated from other Denisovans four hundred thousand and two hundred thousand years ago) to Sapiens populations in the south, interbreeding at a relatively recent date (twenty-five thousand years ago) with Denisovans present in the south.

This genetic can of worms is easiest to understand if we bear in mind that in the Far East, north–south travel was possible on foot during cold periods (as well as hot periods for those who knew how to sail). The same cannot be said for the Far West, where Europe and Africa were separated by the Mediterranean. This was already the case in prehistoric times but it came fully into play after the arrival of Sapiens clans who knew how to cross inlets by boat. And, as the multiple migrations of Oceanians clearly illustrate, Southeast Asia has always been a destination for migrants from the rest of Asia.

The following is an illustration of what is meant by migrations of members of our species in the region: The first *Homo sapiens* arrived some seventy thousand years ago, bringing with them freshly acquired Neanderthal genes from the Near East. Then the ancestors of the Oceanians arrived in the third millennium BCE, followed by the first rice-farming peasants in the second millennium BCE, a number of people fleeing the armies of the Qin emperor at the end of the first millennium BCE, not to mention the many Chinese who, for centuries, founded trading communities in the south. For millennia, these Sapiens migrants brought northern Denisovan DNA from the Asian continent, which was added to that of southern Denisovans already present in the Sapiens genomes of Southeast Asia (particularly among the Negritos and Papuans, whose ancestors rubbed shoulders with the last southern Denisovans probably twenty-five thousand years ago).

Denisovans Lend Sapiens a Helping Hand

Lluís Quintana-Murci's team found that the Denisovan genes that remain in Oceania today contain almost exclusively beneficial mutations that promote the immune response. This means that Denisovan populations of Southeast Asia passed on immune traits to the Oceanian descendants of the first *Homo sapiens* to reach them. These traits favored resistance to the infectious pathogens ubiquitous in tropical rainforests. In 2021, this hypothesis was confirmed by the findings of a team from the University of Aix-Marseille, which included Silvana. The research compared

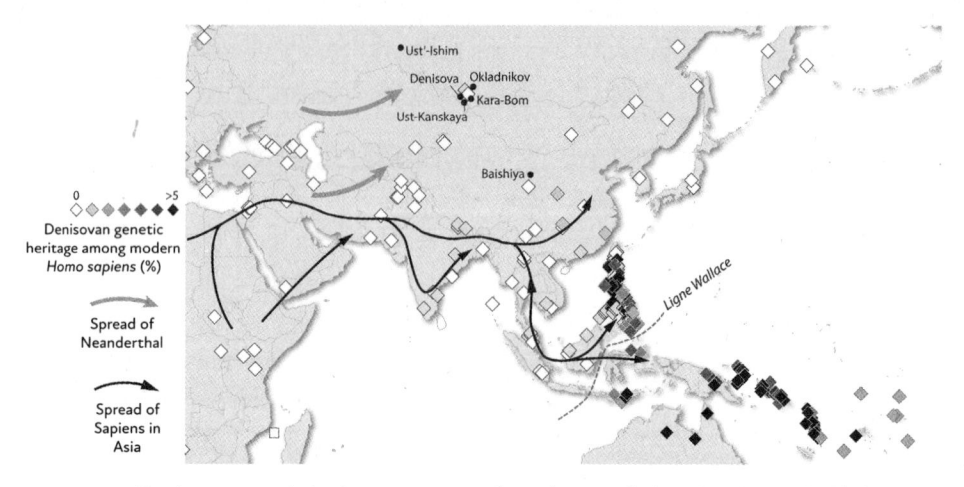

Denisovan genetic heritage among modern-day populations: Denisovan DNA is absent in western Eurasia and Africa (white squares), low in India and continental Asia, and conversely very high in Southeast Asia and Australia. The main sites where *Homo sapiens*, Neanderthal, and Denisova have been discovered are also shown.

Neanderthal and Denisovan blood, whose blood types would have given them a low susceptibility to certain tropical pathogens.

Ultimately, the study of Denisovan DNA from Denisova cave fossils, DNA preserved in sediment, and DNA contained in the genomes of our contemporaries tells us the following: Denisova was an Asian human from the Middle Pleistocene (770,000 to 126,000 years ago, also called Chibanian) who occupied a vast triangle comprising India, Southeast Asia, and continental Asia.

We also know that, like Neanderthals, Denisovans inherited an ancient genetic legacy from so-called super-archaic humans. According to their genomes, Neanderthal and Denisova diverged some four hundred thousand years ago from an African form (*Homo heidelbergensis*) and are closer to each other than they are to Sapiens. But as *Denisova 11* illustrates, they have also mixed a little at the margins of their respective territories, at least in the Altai.

The diversity of *Homo sapiens* living today between Siberia and Papua suggests that, among Denisovans too, phenotypes (the set of observable traits of an organism) were more diverse than genotypes (the set of genetic material of an organism).

However, Denisovan genotypes were sufficiently diverse that northern Denisovans were well adapted to cold and altitude, while southern Denisovans were well adapted to the pathogens of the equatorial forest. The gene distribution of this fossil population suggest that Denisova's ancestors were more numerous in Southeast Asia and southern Asia. Like the Neanderthals, the Denisovans ultimately became extinct after the arrival of *Homo sapiens*.

The facts we've been able to gather on Denisova are certainly valuable, but before any fossils or habitats are successfully identified, we know little about the Denisovans' physical appearance and lifestyles. In the following chapters, we'll draw on a wealth of geological, geographical, climatic, paleontological, and archaeological data to complete the picture we've painted so far. Let's start by getting to know *Homo erectus*, the first and oldest known Asian, and what historical role it might have played.

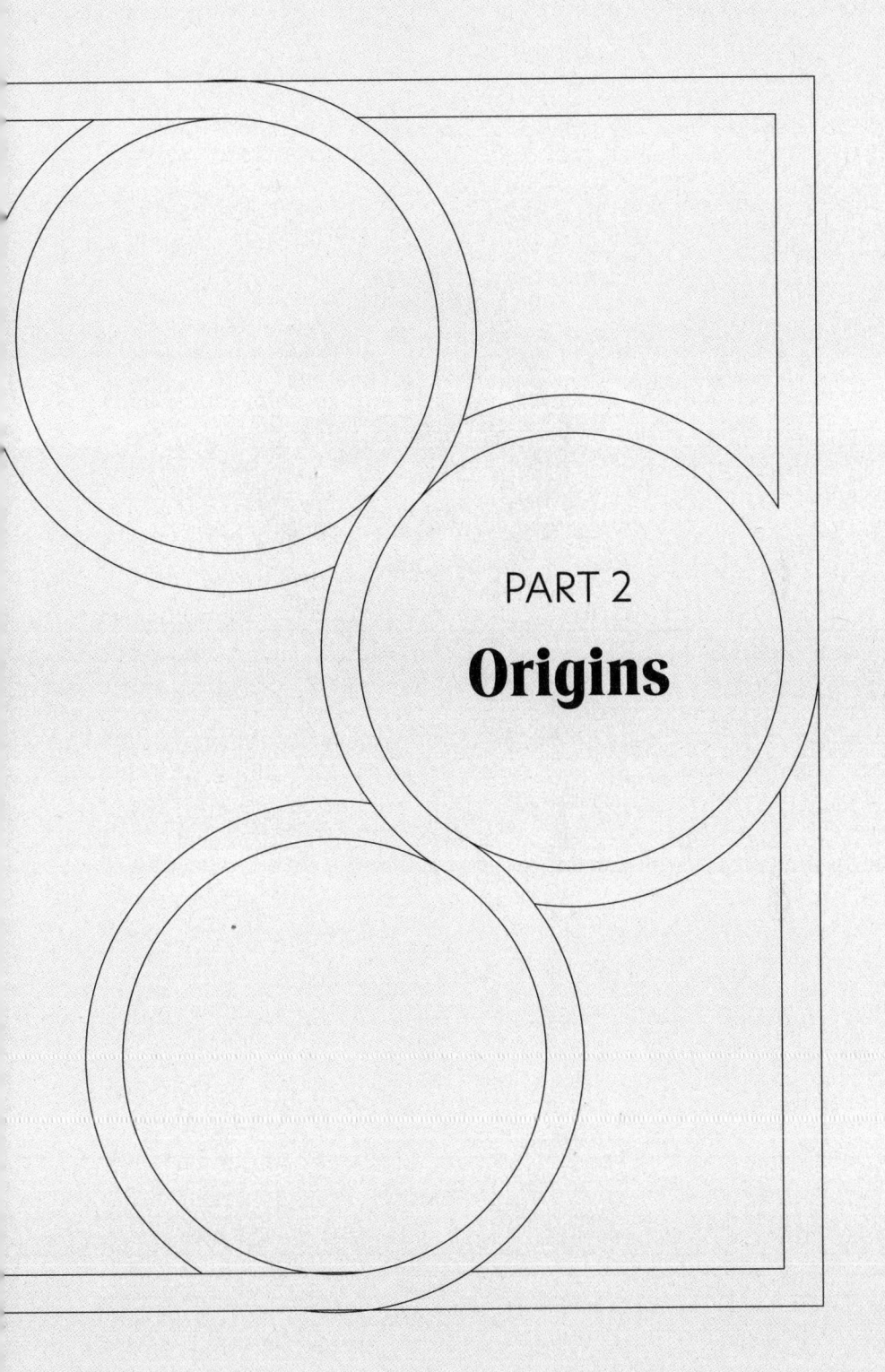

PART 2

Origins

This woman belongs to the second generalist human form *Homo erectus*, which inhabited the entire southern part of Eurasia.

6

Before Denisova: The Curious Case of *Homo erectus*

In 2010, Marine Corporal Richard Bowen made the following revelation on his death bed: "Day after day the war there was getting hotter and closer. . . . The city of Chinwangtao was now under siege by the Communist 8th Route Army with Nationalist gunboats shelling them over our camp. One day a group of them asked us to surrender, saying that they had 250,000 men. To prove the point, that night thousands of fires were lit by them on the adjacent hills and high ground. It looked like Christmas time. From that time on we started digging fox holes at night and napping during the day. I had a 30-caliber machine gun and our lieutenant would, from time to time, change our crossfire. In this nightly digging process we dug a lot of holes. In one of them we found a box that was full of bones. At night it gave us a little scare and we filled in that hole and dug another. Shortly after this we evacuated the area, went back to Tientsin, and then back to the United States with the First Marine Division colors."

This bust of Peking Man was erected in Zhoukoudian in front of the cave where the first skull (now lost) was unearthed in 1929.

Richard Bowen had served at an American base in the Chinese port of Chinwangtao—now Qinhuangdao—and this memory, recounted in his twilight years, was testament to a lost treasure of major importance to China that had been there in the city in 1947: the Peking Man fossils.

These fossils were particularly important to us because they belong to a specialized form of *Homo erectus* known as *Homo erectus pekinensis*, or "Peking Man": the earliest Asian form (though not Denisova's primary ancestor). They were particularly important to the Chinese because they have learned that Peking Man is their ancestor. To understand why, we need to understand the story of the Peking Man, which brings us back to Bowen.

Bowen's son recorded his dying father's words, which were then relayed to Lee Berger at the University of the Witwatersrand, who deemed them credible enough to publish them and inform the Institute of Vertebrate Paleontology and Paleoanthropology (IVPP). In 2012, the year his testimony was published, officials from China's National Cultural Heritage Administration agreed to monitor any work taking place in the area of the former American base, which has since become an industrial wasteland. In 2005, the authorities of Beijing's Fangshan district had already set up a "working committee" tasked with locating the Peking Man fossils. Bowen's recollections were taken very seriously in China because 第一个北京人头骨 ("the first Peking Man skull") is considered the ultimate model of humanity.

Since then, there has been no news on the Peking Man. There is no doubt that, if Bowen's account had yielded a discovery, the news would have caused quite a stir, given that the enigmatic disappearance of these fossils has kept prehistorians on their toes for almost a century. In China, the search for these precious fossils is ongoing; they are suspected either to have remained within the perimeter of the former American base or been hidden in

This photo, taken on January 4, 1941, shows researchers at the site of the discovery of the Peking Man fossils in Zhoukoudian. In the center are the famous German anthropologist Franz Weidenreich and his colleague Pei Wenzhong (front row, third from right), who discovered the first complete skull on December 2, 1929.

a different part of China—or the US—entirely. There are several stories regarding their disappearance, some more fanciful than others. We don't know the truth, but what we do know is the origin of the fossils.

They were discovered over time during international excavations that took place between 1921 and 1937 at the Zhoukoudian site, a series of limestone caves located 26 miles (42 km) southwest of Beijing. One in particular—locality 1—yielded 13 more or less complete skulls, 15 mandibles, 157 teeth, and 11 other bones of the skeleton lost during World War II. We know that the 196 human fossils discovered were first kept at the Peking Union Medical College. In 1941, with the city under Japanese occupation, paleoanthropologists prepared them for shipment to the US.

What happened next remains unclear, but we do know from the testimony of Pei Wenzhong (1904–1982), the famous Chinese paleoanthropologist who discovered one of the skulls, that in late November 1941, just a few days before Pearl Harbor, the bones were transferred to the American Embassy then trucked to the Qinhuangdao base. From there, they were to be loaded onto a US Navy ship that never arrived, as it was captured by the Japanese.

Following Pei Wenzhong's discovery in 1929 of the first partial skull, the "Peking Man," was presented in China—and is likely still considered by many—as the ancestor of modern humanity. The central role of this figure in China places Chinese paleoanthropologists in a difficult position because they have to respect the prestige of this great ancestor. But it is no longer possible to establish precise dates or geological context given that early excavations completely emptied the cave of its fossils, most of which have since disappeared.

First in the Peking Order

Fortunately, the American Museum of Natural History in New York created monographs and casts of the main fossils for the world's leading museums. A tooth discovered at the site by

A page from Franz Weidenreich's monograph on Peking Man,
depicting one of the lost skulls

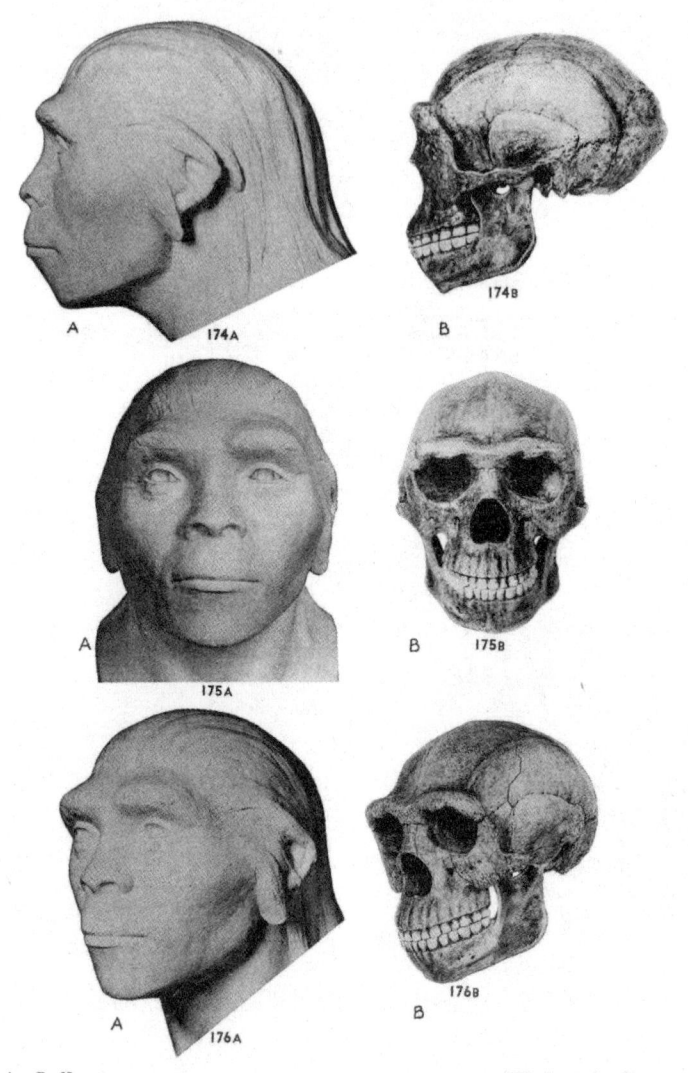

Series D, No. 10

Whole Series No. 127

Franz Weidenreich's physical reconstruction of Peking Man

Swedish paleontologist Carl Wiman (1867–1944), who took part in the excavations in the 1920s, has also been preserved. After having disappeared, this molar resurfaced in 2011 in the collections of the Museum of Evolution in Uppsala.

It appears that *Homo erectus pekinensis* had a cranial capacity of between 850 and 1,225 cubic centimeters; a short stature of around 4.9 feet (1.5 m); robust, thick bones; and strong muscle insertions. This suggests the species was stocky and muscular, well-adapted to the northern conditions of the Nihewan Basin. Its skull also features very strong, rounded, superciliary arches, a carina—a ridge running from the middle of the forehead to the middle of the skull—and an angular neck unlike *Homo sapiens'* rounded neck. The Peking Man also had well-developed temporal ridges, reflecting the working power of its masticatory muscles, made possible by shovel-shaped incisors and strong molars.

How old is this human form? From 1949, after the Communist victory, Chinese paleoanthropologists resumed excavations at the site and completed the process of emptying of the main cave in Zhoukoudian of its archaeological strata, providing a better overview of the cave's sedimentary sequence. Later, the arrival of new dating methods made it possible to analyze the sediment in more detail. By examining the remains of the sediment, Chinese prehistorians were able to reconstruct the 164-foot-thick (50 m) sediment infill of the cave, and to propose a plausible stratigraphic position for each of the fossils discovered before World War II. It was possible to determine with some certainty that locality 1 in the Zhoukoudian cave was occupied between 780,000 and 400,000 years ago.

In 2009, Guanjun Shen's team at Nanjing Normal University confirmed these dates using cosmogenic aluminum-beryllium dating—i.e., the measurement of residual radioactivity of these elements, created on the Earth's surface by cosmic radiation. The Peking Man lived long before the Denisovan era, based on the dates that have been determined from the Denisova and Baishaya caves. Could this be the "super-archaic" human that geneticists have found traces of in the *Denisova 3* genome? If so, then Denisovans interbred with *Homo erectus* before meeting Neanderthal and Sapiens. But where? This is where an important fact comes in: A priori, *Homo erectus* was originally from the south.

Homines erecti in Asia?

The confusion surrounding Peking Man is compounded by a second *Homo erectus* in the mix: Java Man. It was initially given a curious binomial, *Pithecanthropus erectus* by Ernst Haeckel (1834–1919) in his *Natürliche Schöpfungsgeschichte* ("Natural History of Creation," 1868). From the Ancient Greek *píthēkos* ("monkey") and *ánthrōpos* ("human"), the name was coined to describe an intermediate stage between ancestral primates and humans. Dutch physician Eugène Dubois (1858–1940) was fascinated by this idea. Noting that gibbons—monkeys endemic to Southeast Asia—are primates capable of bipedalism, he wondered whether humans might have evolved from some of the tropical monkeys of Southeast Asia. He thought the Dutch Indies, which then covered practically the entirety of Southeast Asia, might be able to provide some evidence. This bold approach was unprecedented for the 1890s. Let's recall that, at the time, the only ancient human fossils to have been discovered were three Neanderthals and the Cro-Magnon fossil, all in Europe.

Dubois set off for Java as a military doctor. During his time there, he began excavating the alluvial deposits of the Solo River at Trinil, near Ngawi, where, in 1891 and 1892, he discovered the first bones. At first, he took these to be chimpanzee bones before the discovery of a skull and femur established them as human. Although Dubois's work was founded on the false hypothesis that the cradle of humanity could in fact be Southeast Asia, it yielded a major discovery: the first ancient human fossil outside of Europe. It was Dubois's determination and the discovery of Neanderthals that led to the creation of the science that attempts to reconstruct the history of the *Homo* genus: paleoanthropology.

H.DANDOY   MAESTRICHT.

Eugène Dubois (1858– 1940), Dutch physician,
anthropologist, and paleoanthropologist: In 1891, in Java,
he discovered the first human fossil outside Europe—the
Java Man, now part of the *Homo erectus* species.

In 1895, Dubois left his colleagues to continue the excavations and brought the precious fossils back to Europe: a cranial vertex, two molars, and a left femur, proving that the species once known as *Pithecanthropus erectus* was indeed bipedal. He received recognition for his fieldwork almost instantly: The University of Amsterdam offered him a professorship, and in 1919 he became a distinguished member of the Royal Netherlands Academy of Arts and Sciences. Despite being showered with honors, after his return Dubois began a long and disappointing struggle to promote his idea that Java Man was an "erect ape-man," the missing link between apes and humans.

After Dubois, Gustav von Koenigswald (1902–1982) made subsequent discoveries of human fossils in Java. With financial support from the Carnegie Foundation, this German paleontologist joined the Dutch Geological Survey of Java and systematically prospected the island between 1931 and 1941. One of his objectives was to determine the age of the Dubois fossils, which to this day have yet to be dated with any certainty. In 1935, he announced the discovery of a child's skull at a site 6 miles (9.7 km) southwest of the town of Mojokerto, in East Java province. He immediately attributed it to the species already found in Java, whose geological age he estimated at around 700,000 years, significantly older than the oldest Peking Man fossils.

In the 2010s, this date was confirmed by a group of Dutch paleoanthropologists, who used the wealth of fauna represented in Dubois's collection to date the layer containing the human fossil. These new investigations make the Java Man between 1 million and 700,000 years old. In 2014, argon–argon radiometry and luminescence dating of sediments contained in engraved shells, presumed to be from the same layer, put their age at between 540,000 and 430,000 years. But how much credence can we give to a collection built up over 120 years ago featuring fossils that were supplied by farmers paid by Dubois for that purpose?

Von Koenigswald was convinced that the fossilized human forms from Java and Europe were related. Continuing his research in Java, he turned his attention between 1936 and 1941 to the Sangiran site in the center of the island. It was here in 1937 that he discovered a first skull fragment, followed by other fossils, before being joined by his compatriot Franz Weidenreich, who had been in China, studying none other than the Peking Man. Following this visit, von Koenigswald and Weidenreich decided to publish their results together, including the discovery of a new skull in 1939.

The characteristics of this skull influenced their decision to group all Asian fossils together—whether "Sinanthropus" or "Pithecanthropus"—under a single term: *Pithecanthropus erectus*.

Margarethe Selenka's excavation in 1907, at the Trinil site, previously
excavated by Eugène Dubois

During World War II, Weidenreich continued to publish the results gathered by von Koenigswald, whom he believed to be
dead. The latter had in fact acquired Dutch nationality and,
having enlisted in the Dutch East Indies army, had been taken
prisoner by the Japanese. Weidenreich finished collecting all the
fossils unearthed in Mojokerto and China under a single species name: *Homo erectus*. This choice was of historic importance
for paleoanthropology. The species name was approved after the
war by von Koenigswald who, having survived captivity, joined
Weidenreich in New York to work together for a period of
eighteen months.

Grouping all the Asian fossils under the same species name
put the researchers in agreement with German American biologist Ernst Mayr, who in 1941, had proposed replacing the
genus *Pithecanthropus* with the genus *Homo*. In the 1940s, the
term wasn't guaranteed to stick, but it took hold, and was even
applied to Africa from the 1970s onward. The idea that this
human form could be generalist enough to be the ancestor of

Franz Weidenreich and Gustav von Koenigswald examining a Javanese fossil
at the American Museum of National History in New York.

subsequent Asian populations became widespread, to the point where it was taken as a given that *Homo erectus* had been the only human in Asia.

Meteor Alert

This idea still prevails, and yet it's not true. Since the idea of the "absolute reference of humanity" took hold in Asia, the discovery of numerous other fossils in the same age range in Java, mainland China, and western Eurasia has complicated the vision of those pioneers of paleoanthropology (Eugène Dubois, Pei Wenzhong, Franz Weidenreich, and Gustav von Koenigswald). Clarification of the ages of the fossils and their geological and climatic context has, over time, led researchers to conclude that there were two or even more species in Asia, which may either have evolved simultaneously or at different times, in different climates and in different places.

The chronology initially proposed has also changed as more accurate dates have been established for fossils from Java's three main ancient Paleolithic sites (Trinil, Sangiran, and Mojokerto). In Java alone, researchers have been able to distinguish two major groups from the fossils discovered: one more recent group and a second archaic group (which is the one we are most interested in here). By cross-referencing all available chronological clues, Shuji Matsu'ura's team at the National Museum of Nature and Science in Tsukuba, Japan, has succeeded in dating the arrival in Java of Eugène Dubois's *Homo erectus* (in Trinil) to around 1.3 million years ago, making the Java Man roughly half a million years older than the Peking Man.

During this half-million-year period, the fall of a mega meteorite in the China Sea region must have caused a split between pre- and post-catastrophe Asian populations. The impact of this meteorite may well have wiped out entire populations in Southeast Asia and even in those living in the south of the Asian continent. During the half-million years of evolutionary time separating Peking Man from Java Man, new forms of human biology and cognition emerged in Africa and spread to Eurasia, replacing the extinct Southeast Asian populations. Grouping together the Peking Man and Java Man is tantamount to merging two forms at different evolutionary stages, living in very different environments at very different times.

Yet among paleoanthropologists, stating that the two may be different species makes about as many waves as a mega meteorite. The idea has taken root because their anatomies appear to suggest they could be one and the same. There are certainly similarities between the bone characteristics of Peking Man and Java Man: Their skulls both have a small volume—900 cubic centimeters—a high bone density, wide masticatory muscle attachment points, pronounced superciliary arches, robust faces with developed cheekbones, strong mandibles, and so on. However, the fact that they share certain "archaic" features—i.e., those inherited from an ancestral form—does not necessarily indicate that the two belong to the same species.

The two human forms also display "derived" traits specific to their lineages, acquired as each evolved respectively. For example, the molars of both Peking Man and Java Man are large, but those of the former are much larger with a different cusp formation. The extremely marked macrodontia of *Homo erectus* from Java can actually be explained by hundreds of thousands of years of chewing tough plants known as "C4 plants" which are highly nutritious and ubiquitous in Southeast Asia. If one thing is for certain, it's that nothing is simple in paleoanthropology. It's possible for certain similar derived traits to appear in two distinct lineages, making anatomical comparisons difficult.

A Science Influenced by Genocide

All this highlights the uncertainty surrounding the concept of an Asian *Homo erectus*, which in turn makes it difficult to determine its relationship to Denisova. Since the two human forms Peking

Man and Java Man lived on opposite sides of a cataclysm, in very different climates, 3,400 miles (5,500 km) and half a million years apart, they really can't be the considered one and the same—so why does this continue to be the case?

To understand what has happened, we need only look back over the first half of the twentieth century. First of all, evolution was considered to be a linear progression, rather than a series of many different branches. Due to a lack of fossils, there was no way to determine with any certainty either the age of the *Homo* genus or the speed at which new human forms were evolving. What's more, the idea that several species could have lived at the same time was unthinkable, when evolution was perceived as having a "goal": *Homo sapiens*.

It's also worth considering the context in which Franz Weidenreich was making these discoveries. He had arrived in what was then known as Peking in 1934 at the age of sixty-one, having lost the German Alsace of his youth and the university post he held there as a town councilor for Strasbourg, seen the rise of Nazi ideology in Germany, and then again lost his post at the University of Heidelberg—this time because he was Jewish. Then, three years after his arrival in China, he witnessed the Nanjing massacre, one of the worst wartime atrocities committed in the twentieth century against a civilian population who were considered "inferior" in the archetypal ideology of Japanese "superiority" over other Asians.

It's easy to see how this great researcher's scientific and humanist convictions might have played a part in the development of his theory. In his words: "The Sinanthropus's [Peking Man's] upper incisors have a distinctive 'shovel' shape. This peculiarity is found in the corresponding teeth of the modern Mongolian race. This, along with its *torus mandibularis* appears to prove that the Sinanthropus has its place in the lineage leading directly to the modern human and that, in present-day mankind, the Mongols are those most closely related to Peking Man."

This statement, and the mention of a human "race," seem outdated to us today, but these are in fact the basis for the contemporary understanding of *Homo erectus*. Weidenreich's intention was to draw upon the observable evidence that, from a biological point of view, all Sapiens populations are interfertile and belong to a single species distributed throughout the planet's ecosystems. In the midst of the bloodthirsty madness of the 1930s and '40s, his remark reflected not only the very scientific need to be cautious, but also the humanist rejection of the "race hierarchy" that underpinned Nazism, Japanese nationalism, and various other forms of racism. In a lecture published just two years before his death, he declared: "All available facts indicate that crossbreeding is not a late human acquisition that took place only when Man reached his modern phase, but that it must have been practiced from the very beginning of Man's evolution."

Franz Weidenreich was in fact putting forward a scientific theory that ran counter to the prevailing racism. According to his doctrine, *Homo sapiens* from all the major eco-regions of Africa and Eurasia evolved in parallel, exchanging genes from one region to the next, so that the human gene pool remained unified with a few different nuances. Even if the Mongolian, European, African, and other "races" appear distinct, they all belong to one and the same species. In 1984, this theory inspired a second, more elaborate theory along the same lines.

That year, Milford Wolpoff of the University of Michigan, Alan Thorne of the University of Sydney, and Xinzhi Wu (1928–2021), Vice Director of the IVPP, proposed an alternative to the dominant *Homo sapiens* "replacement theory": the "multiregional theory." According to this doctrine, all human populations evolved across the different regions of the world in parallel toward *Homo sapiens*, and the unity of our species was established and maintained through migrations. Xinzhi Wu expressed this idea by coining the concept of "continuity [of evolution] with hybridization." While the multiregional theory was clearly based on Weidenreich's ideas, it gave a scientific basis to a deep-rooted tendency in China toward "racial nationalism."

According to a point of view that has long prevailed in China, and no doubt persists to this day, the origins of modern humans are local. For the Chinese, this hypothesis presupposes a foregone conclusion: There is a Chinese archetype, which has been maintained throughout time since the Peking Man. Sinologist Barry Sautman from the Hong Kong University of Science and Technology, has studied this ideology closely: "Nowhere is this phenomenon more pronounced than in China, where these disciplines form the canvas of Chinese 'racial' nationalism. This nationalism holds that each of us can trace our identity back to a biologically and culturally distinct community, whose 'essence' has been maintained over time. It differs from ethnic nationalism only in that it biologizes and extends group identity by adding a racial component to more sharply distinguish the nation from its neighbors."

This trend is reminiscent of that which dominated France toward the end of the nineteenth century. Prehistory was instrumentalized to such an extent that in France, just three Sapiens fossils—Cro-Magnon, Grimaldi, and Chancelade—were used as a basis to determine the origins of the human "races" that were believed to exist at the time: black, white, and yellow. When Europeans introduced prehistoric science to China in the early twentieth century, scientists in the Middle Kingdom also succumbed to the delicious temptation of believing themselves to be at the center of the world.

Denisova in the Shadow of *Homo erectus*

The ideological role prehistory plays in China influences both Chinese prehistorians and the Westerners who work with them. These scholars have created a curious vocabulary to communicate these ideas. They refer to the "reference" form of Peking Man as *Homo erectus* sensu stricto ("in the strict sense") while

European and African forms of *Homo erectus* are referred to as *Homo erectus* sensu lato ("in the broad sense").

The issue here is that the oldest known *Homo erectus* fossils found in Africa (see the box on pages 101–3) date back 2 million years, making them up to 1.2 million years older than the Peking Man. How is it possible that *Homo erectus* "in the broad sense" predates *Homo erectus* "in the strict sense"? This dichotomy creates a false pretense, covertly reversing the direction of evolution and going against the most elementary scientific common sense: In reality, evolution moved from Africa toward Eurasia, rather than in the opposite direction.

What's more, forms more evolved than the Peking Man are referred to as "evolved" *Homo erectus*, or, if they differ greatly from it and share derived traits with our own species, "archaic" *Homo sapiens*. This is the case, for example, of the "Dali Man," a Chinese skull around 250,000 years old. From the outset, the paleoanthropologists who studied Dali Man recognized that it differed from the fossils in the same classification group as Peking Man, the most recent of which date back 400,000 years. They considered Dali Man first as a "late" *Homo erectus*, then— like their mentor Weidenreich—as a transitional form between Peking Man and *Homo sapiens*, heralding the emergence in Asia of a "mongoloid" form of *Homo sapiens*.

The truth is that prehistorians have had to adapt to a shocking reality: Our vision of the prehistoric settlement of Asia is in total disarray. There's a glaring oversight: *Homo erectus* is supposed to have populated Asia without interruption for 1.3 million years (i.e., for about half the time of the *Homo* genus's entire existence) and for five times as long as *Homo sapiens*. How could a single species, *Homo erectus*, prevail in such an immense territory, with such different climates and environments, evolving so little over such a long period of time? In Europe, this was not the case. We know that new forms from Africa arrived, took over local populations, developed distinct characteristics and eventually came to dominate.

The sense of unease this dissymmetry between western and eastern Eurasia caused researchers is reflected in the lexical complexity that characterizes all fossils classified as *Homo erectus*: *Homo erectus erectus, Homo erectus pekinensis, Homo erectus mauritanicus, Homo erectus modjokertensis, Homo erectus leakeyi, Homo erectus ngandongensis, Homo erectus yanoumensis, Homo erectus bilzingslebensis, Homo erectus tautavelensis,* "archaic" *Homo erectus, Homo erectus sensu stricto, Homo erectus sensu lato,* "evolved" *Homo erectus,* and even *Homo sapiens erectus.*

In Asia, the same name—*Homo erectus*—is used to describe all the forms that existed between 1.3 million and 100,000 years ago, from the oldest *Homo erectus* in Asia to Denisova. This is vague, false, and impedes scientific thought. This view of Asian settlement is also at odds with the earliest known dates of each of the successive human forms and their progressive expansion into Eurasia. In this context, the appearance of Denisova has thrown sand in the gears for those attempting to Sinicize the origins of humanity.

What is the solution? The only solution is to come back to where it all began: Africa. In order to situate Denisova in the grand scheme of human history, we have to understand the successive waves of humans leaving the cradle of humanity to conquer Eurasia.

Homo erectus, a Two-Million-Year-Old African

The super-archaic human that Denisova encountered in Asia actually came from Africa, as did all human forms that ever entered Eurasia. In 1971, at the Koobi Fora site near Lake Turkana in Kenya, Kenyan paleoanthropologist Richard Leakey (1944–2022) discovered KNM-ER 992. The name might sound like it belongs to an extraterrestrial, but it is in fact a large mandible, belonging to the first "African *Homo erectus*" fossil dating back 1.5 million years. In 1984, his collaborator Bernard Ngeneo unearthed an even older skull on the same site, dated at 1.7 million years: Its features, notably a massive supraorbital ridge and receding forehead, are similar to those of Asian *Homo erectus*.

Perhaps the most impressive "African *Homo erectus*" was found in 1984 in Nariokotome, Kenya, by Kamoya Kimeu, another of Richard Leakey's collaborators. The specimen was named "Turkana Boy." The skeleton remains practically intact and is 1.6 to 1.5 million years old. In 2020, an international team discovered the skull of a 2- to 3-year-old "ancient *Homo erectus*" child, 1.9 million years old, in South Africa's Drimolen cave. In 2023, Margherita Mussi's team at Rome's La Sapienza University redated another infant mandible fragment, discovered at the Garba IV site in Ethiopia, to over 2 million years old. Silvana had determined in 2003 that the specimen belonged to *Homo erectus*. These fossils, all found with tools characteristic of *Homo erectus*, suggest that African *Homo erectus*

The Kenyan paleoanthropologist Richard Leakey (1944–2022) comparing the skulls of two hominids: We owe some of the major fossil discoveries in east Africa, and particularly in Kenya, to Leakey's team.

emerged before 2 million years ago—i.e., at least half a million years before the oldest *Homo erectus* in Java.

The eminently erectus features of the Turkana Boy and other *Homo erectus* fossils, as well as their ages predating Asian fossils, initially bothered paleoanthropologists, who proposed to use the name *Homo ergaster* ("human craftsman") to designate the African *Homo erectus*, but usage of the term seems to be in decline. In any case, we know that *Homo erectus* was already living in the cradle of humanity probably 2 million years ago, where it remained until up to 1 million years ago—that is, at least 600,000 years before the *Homo erectus* of Java. This is hardly surprising; we already know from the Ubeidiya site in Israel that *Homo erectus* emerged from Africa, then, around 1.5 million years ago, progressed toward Asia and Europe. A mandible and a fragment of a face dating back 1.4 million years found at Sima del Elefante in Spain attest to the arrival of the African human form in the otherwise mild climate of southern Europe.

The lion is one of the great species found in both Asia and Africa. There's no doubt that humans kept tabs on lions to scavenge left-behind carcasses, which they would have had to fight the hyenas for.

The Denisova cave: a luxurious mountain apartment from a prehistorian's point of view. Its entrance can be found on a cliffside in the Sibiryachikha valley in the Altai.

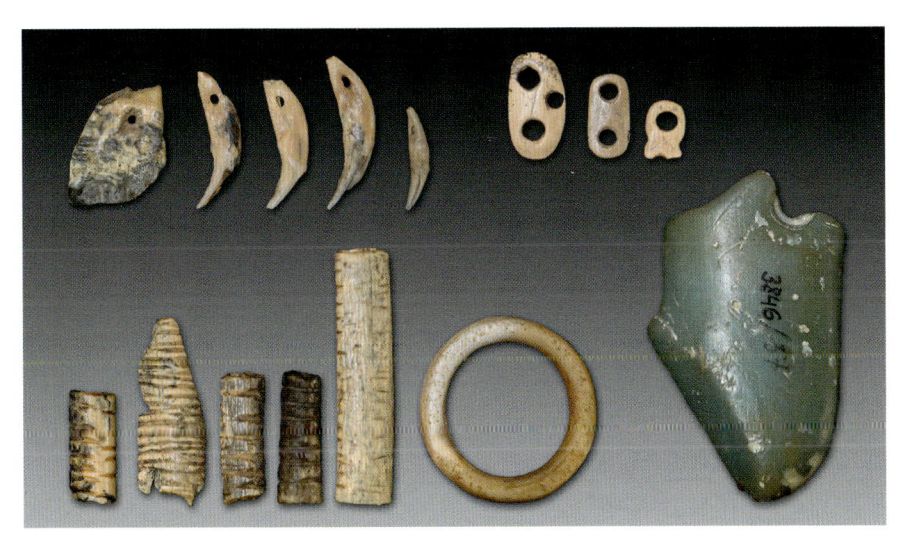

The objects discovered in the upper levels of the cave are astonishing: beads made of bone and deer spit, a fragment of a chlorite bracelet, and various objects of unknown use made out of bones with stripes carved into them. These objects date back forty to fifty thousand years, to the transition period between Denisova and Sapiens in the Altai.

A group of researchers examines the Denisova sedimentary deposits, enabling them to go back in time more than two hundred thousand years.

On the island of Flores in Indonesia, south of the Wallace Line, this cave has yielded the remains of a strange, specialized, diminutive human form: the Flores human.

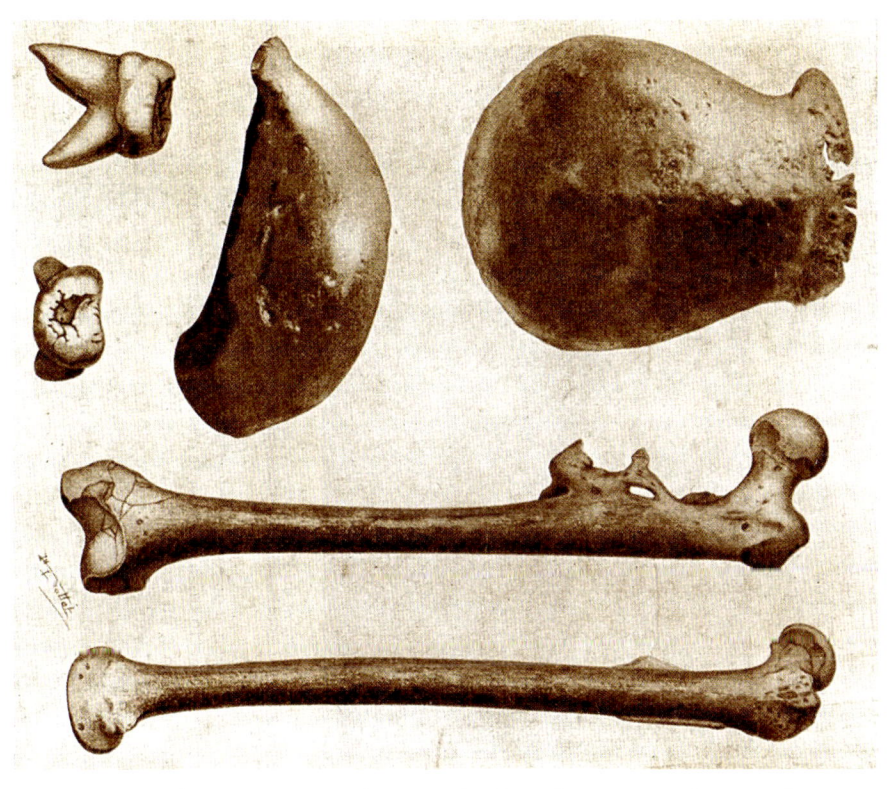

The cranial vertex, femur, and molar of *Pithecanthropus erectus*, discovered in Java by Eugène Dubois at the end of the nineteenth century, were the first human fossils to be discovered in Asia.

The Baishiya Karst Cave, where the Xiahe mandible was discovered, is located in Gansu province, central China. It was a monk meditating in this holy place who spotted the fossil in the depths of the cave.

An excavation team from Lanzhou University analyzing the strata of the Baishiya Karst Cave. As the cave is situated at an altitude of 10,764 feet (3,281 m), the researchers are working in very cold conditions.

Homo rhodesiensis/heidelbergensis (top right, known as the "Broken Hill" skull) shares archaic features with other fossils, including *Homo erectus* ("Sangiran 17," top left) and Neanderthal ("La Ferrassie 1," bottom left) but has little in common with its latest descendant, *Homo sapiens* (bottom right).

While *Homo erectus* from Java ("Sangiran 17," left) and Neanderthal ("La Ferrassie 1," right) have skulls that are elongated towards the rear, that of Sapiens (center) expands in height.

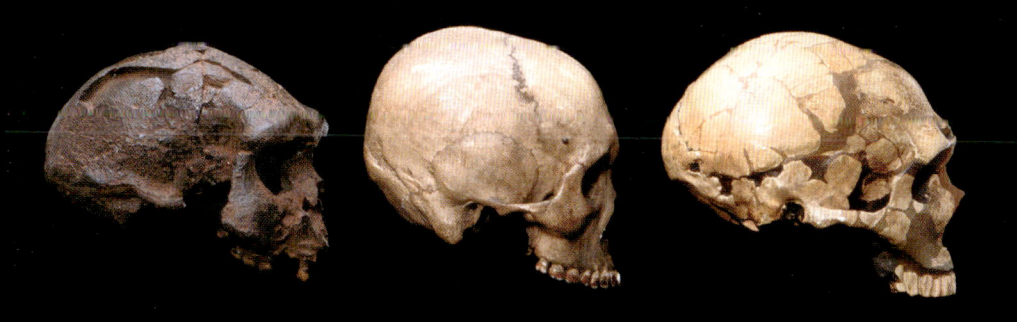

Found in 1867 in the commune of Abbeville, this tool is the perfect example of a sophisticated hand axe fashioned by *Homo heidelbergensis* and its descendant, Neanderthal.

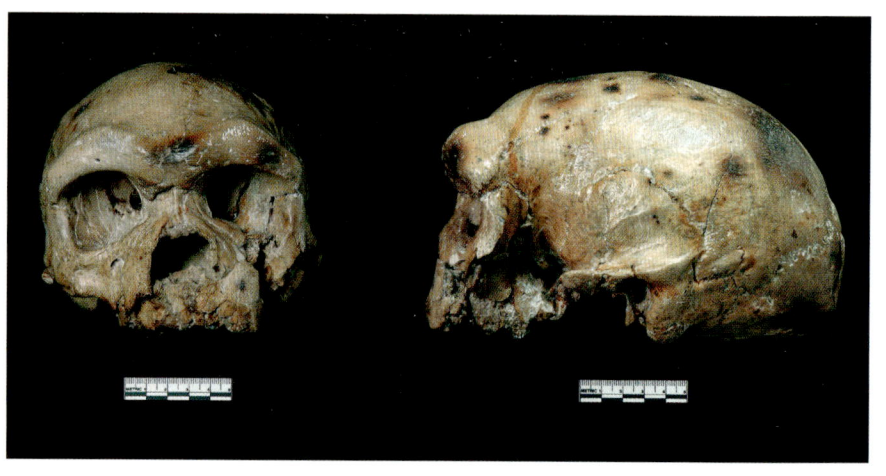

The Dali skull was discovered in 1978 and has several typically Denisovan features, such as thick, discontinuous superciliary arches and a very broad face. The first Denisovan?

The Harbin skull is the best preserved of all Denisovan fossils. Discovered in 1933, during the Japanese occupation, it comes from a bank of the Songhua River not far from the city of Harbin, the capital of Heilongjiang, China.

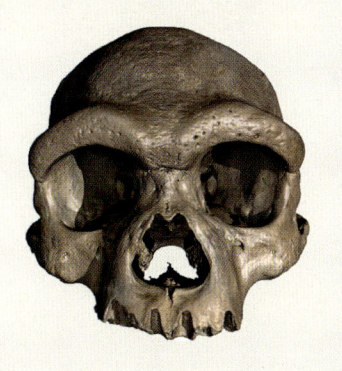 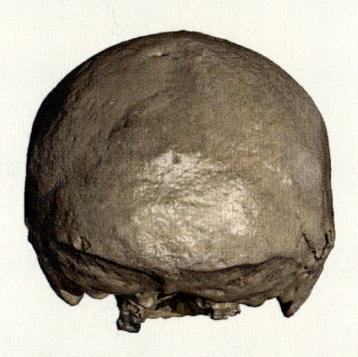

50 mm

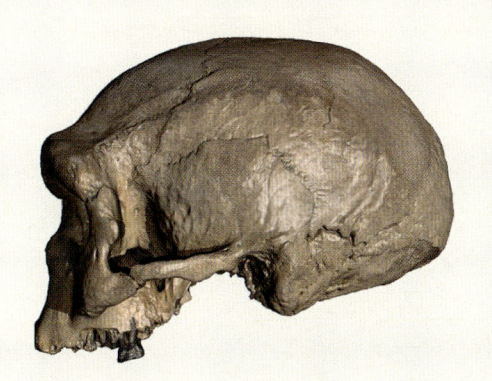 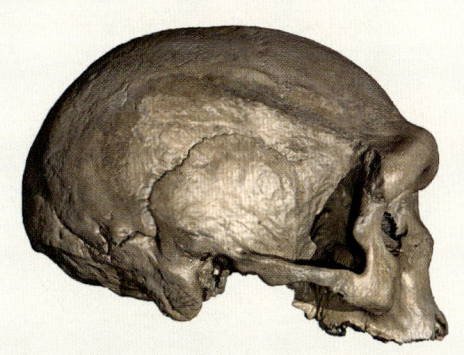

 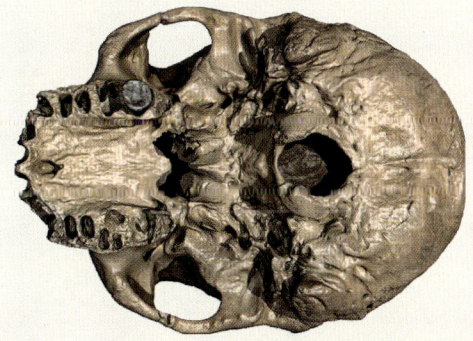

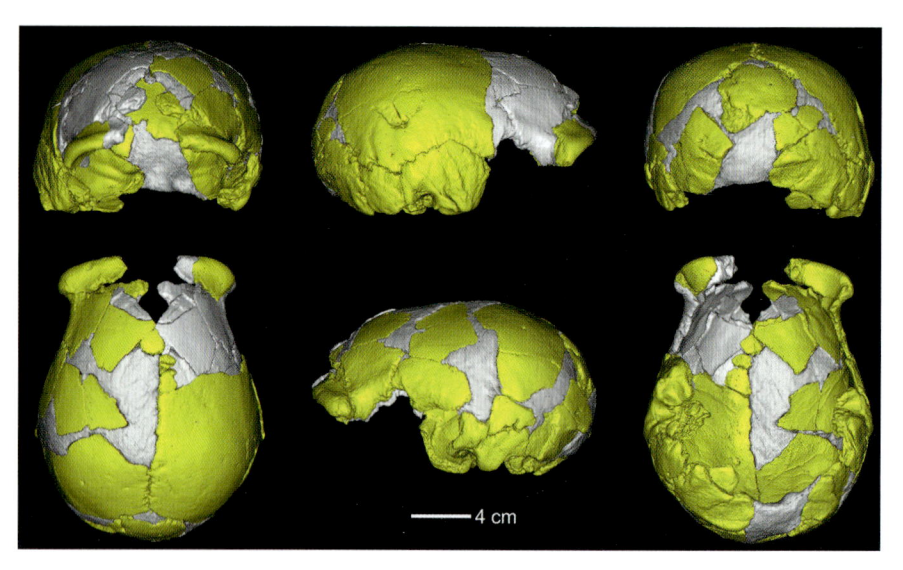

The Lingjing skulls have several characteristic Denisovan features, such as a football ball shape and large endocranial volume, as in Neanderthal, as well as a supraorbital bulge that appears to protrude to the sides.

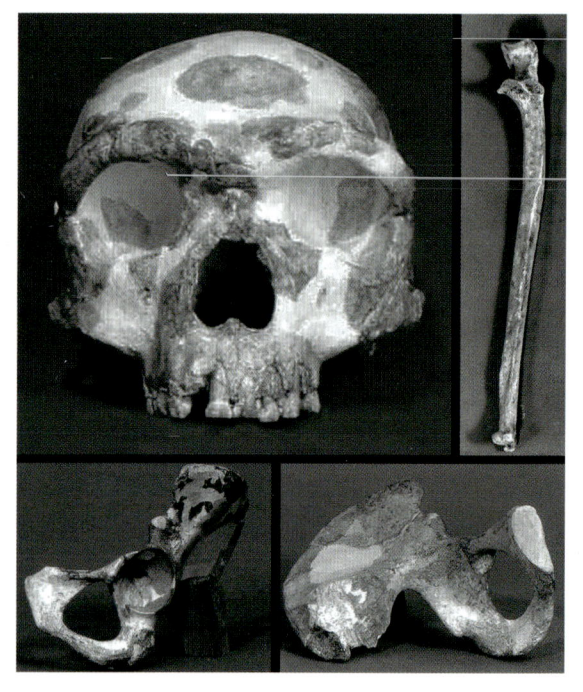

Although only partial, the skeleton of the Jinniushan human gives us information about the person's height, weight, and sex: She's a woman. This is the only Denisovan skeleton that has ever been discovered.

7

Earth Systems: The Science of Denisova

D id Denisovans evolve on a continent that no longer exists? Yes. We know from genetics that most of them lived in Southeast Asia, a region that was periodically submerged or only partially above sea level. A study by Harold Voris, of the Field Museum of Natural History in Chicago, shows that for 40 percent of the last 250,000 years, the part of the Sunda continental shelf that is now a little less than 130 feet (40 m) deep was above water. For most of the last 2 million years, the surface of the world's oceans was well over 130 feet lower than today's levels. We can therefore conclude that Denisova spent most of its evolution living on an immense continent constituting the majority of the Sunda Shelf, Malaysia, Sumatra, Borneo, Java, and continental Asia.

This continent, known as Sundaland, extended as far as the Wallace Line, a deepwater channel and biogeographical boundary separating two types of fauna. This line separates Sunda from a large tropical continent to the south. This continent was the main destination of humans migrating from Africa, since they tended to gravitate toward climates to which their bodies were well suited, and Denisova's ancestors populated and reproduced across the continent of Sundaland. During hot periods, the continent became a fragmented archipelago much like today, and populations taking refuge on the large islands drifted genetically.

In short, Denisovans evolved from an ancestral population that had settled in the tropical south of continental Asia and the immense continent of Sundaland. How and why did they end up there? To grasp this fully, we need to understand why, ever since humans began migrating from Africa across Eurasia, the Earth's biogeographic system has forced them to follow much the same migratory trajectory.

Earth System Science

Earth system science is a discipline aimed at integrating our knowledge of how our planet functions as a system, whether in its geological, climatic, or biogeographical aspects. This can help us to understand the settlement of Asia since the emergence of the *Homo* genus.

Let's begin with a crucial fact: For some three million years, the distribution of continental masses on the Earth's surface has remained more or less the same. The Earth comprises three main climatic zones within each hemisphere: polar/subarctic, temperate, and subtropical. Over the last three million years, the climate in these zones has remained largely stable, with just one exception.

And it's an important exception. On a geological scale, the ocean continually rises and falls as glacial episodes shift the boundaries southward between climatic zones. There are three of these zones (also known as ecozones): arctic and subarctic, temperate, and tropical (see page 113). Over the last three million years— the period corresponding to humanity's evolution—glacial cycles have shifted the boundaries of these ecozones and changed the volume of the landmasses submerged. These so-called plio-quaternary glacial sequences consist of a series of at least twenty-five

Map 1

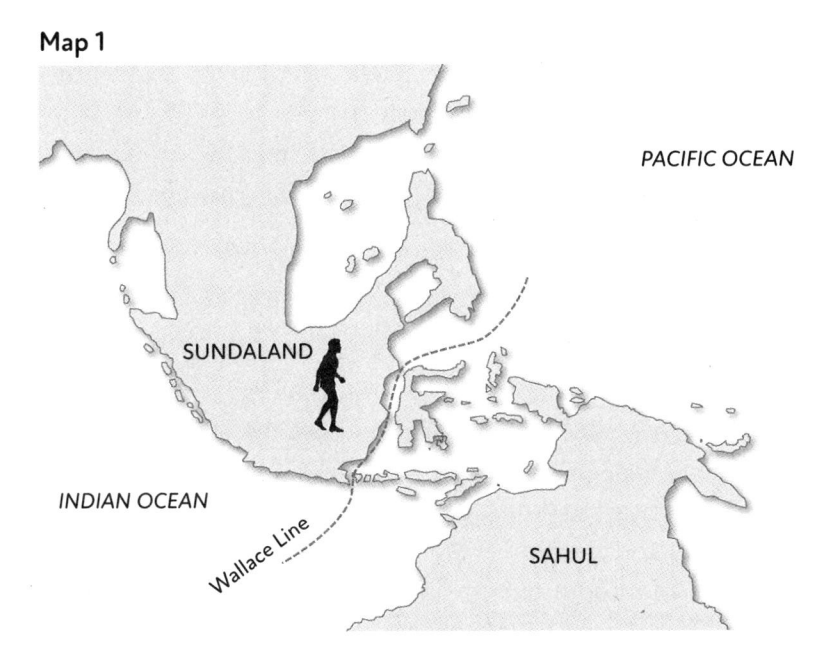

Map 2

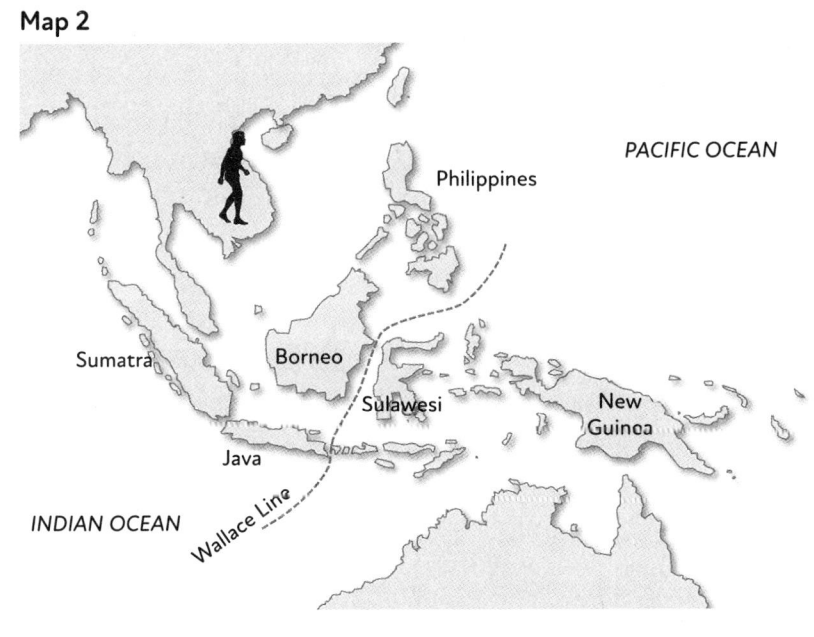

Sundaland, the Philippines, Sulawesi, New Guinea, Sumatra, Borneo, Java, and
continental Asia during the Ice Ages (1) and interglacial period (2)

cycles, consisting of an interglacial followed by a large-scale freezing over of northern Eurasia. Some eight hundred thousand years ago, the length of these cycles increased from forty thousand to one hundred thousand years, a point we'll come back to later, as this played a crucial role in the arrival of Denisova's ancestors in Asia.

Ice Ice Baby

During the glacial maximums—i.e., the coldest few millennia of the cycle—ice sheets (giant glaciers that can be miles thick) eroded many of the plains of northern Eurasia and the ocean was 160 to 460 feet (50 to 140 m) lower than today. For this reason, many parts of today's submerged continental shelves were then above water and populated by humans, starting with the Sunda shelf—i.e., Southeast Asia. At the start of the interglacial periods (between glacial maxima), the same ice sheets melted, and their water and sediment were channeled into rivers running southward during periods of low oceanic water.

Then, as the waters gradually rose, they extended southward in inner Eurasia (Danube, Volga, Dnieper), westward toward the outskirts of Eurasia (Garonne, Loire, Seine, Somme, Rhine, Elbe), northward (Yenisei, Taimyr), and eastward (Thames, Amur, Tumen, Yellow River, Yangtze River). These rivers cut through the mountains, creating immense alluvium valleys and vast sedimentary basins ending in lush deltas. Throughout history, these valleys, basins, and deltas have provided generous habitats for large herbivores and their predators, including hunter-gatherers.

Following each of the ice ages, the climate became more temperate farther toward the north. Lush habitats including the Eurasian Steppe formed around these great rivers to the south of the subarctic forest. This macro-ecosystem of temperate grasslands and shrublands, situated around the fortieth parallel north, formed a steppe corridor linking the heart of Europe (Hungary, Romania) to Northern China (Inner Mongolia). The corridor, with its large herbivores multiplying in their millions, was essentially an immense larder for hunter-gatherers. These mobile resources were easy to spot along the steppe, attracting hunter-gatherers of all eras who traveled transcontinental distances. This explains how Neanderthals and Denisovans met in the Altai. The steppe was something of a highway that has always connected eastern and western Eurasia. The climate conditions for hunter-gatherers living along this highway in both directions were continental: very cold in winter and very hot in summer.

Great Subtropical Variations

In Africa, subtropical and tropical climates extend from the Sahel to the deserts of southern Africa. In Asia, these climates can be found from the south of the Arabian Peninsula, through India and all the lands south of the Himalayas, to the Indochinese Peninsula. In Southeast Asia, they are found on the Sunda Shelf, and beyond to Northern Australia. Finally, in continental Asia, they exist in the Qinling Mountains, a 930-mile (1,500-kilometer) east–west mountain range spanning modern-day China, which marks the northern limit of areas affected by the monsoon (see ecozone map, page 115, or Chinese geography, page 175).

To reach Asia, prehistoric peoples had to leave Africa. They could have done so via the Levant, but it's also possible that during certain periods, they arrived via the Bab-el-Mandeb strait separating eastern Africa from the southern Arabian Peninsula, which has varied in depth over hundreds of thousands of years due to tectonic shifts and changing sea levels. It's still safer to assume that the strait was largely inaccessible to humans, unlike the Levant that would have led them to the Arabian Peninsula. During wet periods, the monsoon advanced as far as the Levant, creating constellations of lakes that would have made the peninsula a pleasant route for traveling to Asia.

This had two essential consequences: The first is that African fauna, including humans, were able to reach the Levant via the

Nile valley, then settle in the Arabian Peninsula. Secondly, during glacial periods, the world's oceans dropped for very long periods by more than 160 feet (50 m). This led to the Persian Gulf partially drying up (its average depth is 160 feet) extending Mesopotamia southward and facilitating the passage of tropical fauna from the Levant and the Arabian Peninsula eastward. They then spread along the coasts of what are now Iran and Pakistan to reach the Indian subcontinent.

The Earth's Great Pressure Cooker

In the Earth system model described above, Africa is like a great pressure cooker, intermittently spouting surges of humans like puffs of hot air toward Eurasia. During the course of each human wave into Eurasia, the following scenario repeated itself:

1. Humans left Africa via the Levant and/or Arabian Peninsula.
2. They ventured along the southern Asian shore to the Wallace Line in Southeast Asia, and to the Qinling mountains in continental Asia.
3. They stopped off in western and eastern Eurasia, the gateway to the colder north.
4. They would then interbreed and engage in cultural exchange with members of preceding waves, which eventually led to the acquisition of cold adaptation techniques.
5. Once they had mastered fire, it became possible to progress northward to higher latitudes, as they had effective protection from the cold.
6. Circulation toward the north between western and eastern Eurasia via the steppe highway connecting the heart of Europe to continental Asia meant western populations would come face-to-face with eastern populations.

We believe this process repeated itself several times, with four well-identified major human waves coming from Africa (and perhaps other prehuman waves before that), which correspond to the four successive evolutionary stages of the *Homo* genus that produced clearly distinct forms. Let's take a brief look at some of them.

The first, which we believe to have emerged earlier than 2 million years ago, corresponds to a form close to that of the African *Homo habilis*. Although its presence in Eurasia is archaeologically uncertain, comparable skeletons have been unearthed at the 1.8-million-year-old site of Dmanisi, Georgia—the gateway to Europe—suggesting that this distant ancestor had ventured into Eurasia in a warm climate, following the African fauna that had taken the same trajectory.

The second form that left Africa around 1.5 million years ago is *Homo erectus*, which we've heard a lot about already, since it's so often wrongly confused with the wave that included Denisova's ancestors.

The third form, which was part of the wave that left Africa around eight hundred thousand years ago, is that of *Homo rhodesiensis/heidelbergensis*. As we saw in chapter 4, genetic data suggest that this form is the origin of Neanderthal in Europe and Denisova in Asia.

The fourth wave is *Homo sapiens*, which up until very recently, has been the easiest to trace. We know that this migratory wave began early, perhaps as early as two hundred thousand years ago, but genetic data indicates that the main part of it—the one that left its legacy in all modern humans—began around seventy thousand years ago or later.

All of these waves spread successively across Eurasia at increasingly shorter intervals, following different dynamics depending on the respective levels of cultural evolution of each population. While the first two waves who had not domesticated fire were limited to fairly warm regions, the third would move into temperate and subarctic regions, while the fourth—ours—would

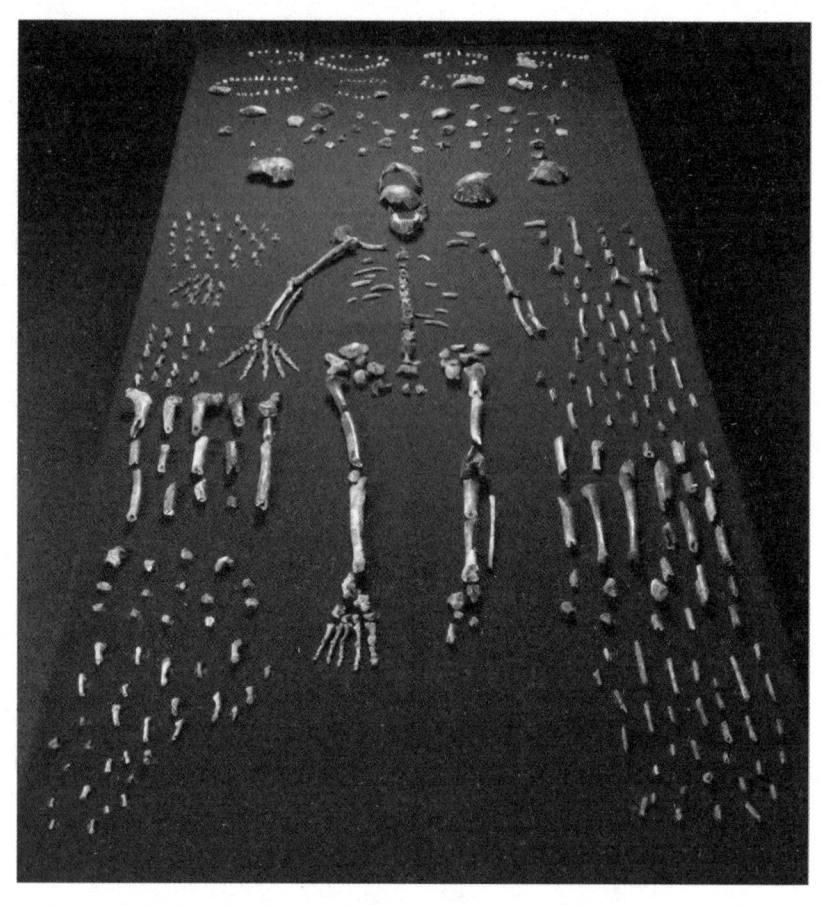

Homo naledi, whose fossils were discovered in a network of caves in South Africa, is an example of a specialist human form. This small human (4.9 feet/1.5 m tall) lived some 300,000 years ago at the same time as the first *Homo sapiens* and was adapted to a lifestyle that required frequently climbing trees and rocks.

reach as far as the North and South Poles. We call the human forms associated with these waves "generalist" because, instead of locking themselves into a single ecosystem as some "specialist" human species have done, in the course of their evolution they have taken over all the ecosystems of the Old World. The generalist human forms from each evolutionary era brought a certain bone anatomy (general form and the biology and cognition associated with it) to Eurasia, as well as material cultures (tools and activities) that can be traced back to Africa first.

This process was repeated for two million years. It also applies to the case of *Homo erectus*, suggesting an African form that evolved in the cradle of humanity at least half a million years before daring to set foot in Eurasia. Similarly, we know that *Homo sapiens* could have been a pan-African species for at least three hundred thousand years and began to spread across Eurasia two hundred thousand years ago. Our species then took over this continent, leading to the complete extinction of other human forms only forty thousand years ago. Sapiens finally reached all corners of the globe on February 19, 1819, after Captain William Smith sighted the last unexplored continent, Antarctica, from his ship *The Williams*. Denisova must have come to settle in Asia in much the same way. Now let's take a closer look at how the Earth system produced Denisova, as well its genetic and archaeological impact on the species.

The Earth's Three Ecozones

Each of the planet's hemispheres contains three climatic bands known as ecozones. We're going to take a look at those in the Northern Hemisphere, which prehistoric peoples crossed. The first ecozone is made up of the polar and subarctic lands, which form a vast biogeographical region composed of tundra, forest tundra, taiga (boreal forest), peat bogs, and, to the south, mixed forests (containing conifers and a few deciduous trees). They extend in a fluctuating pattern from Norway (sixty-fifth parallel) in western Eurasia to central Siberia, and from there to northern Kamchatka in eastern Eurasia (sixtieth parallel).

The second ecozone encompasses today's temperate lands, which currently include vast areas of Europe as far north as Scandinavia (sixtieth parallel, Oslo latitude), central Asia north of the Taklamakan and Gobi deserts, and Northern China from the latitude of the Qinling Mountains (thirty-fourth parallel) to Inner Mongolia and beyond, to halfway across Eastern Siberia (sixtieth parallel). The boundary latitudes between subarctic and temperate zones vary depending on sea currents and the contours of the land, which fluctuate based on sea levels. In the center of the continent, in the Altai (where the Denisova Cave is located), the climate is very much continental, which means it can be very hot in summer and very cold in winter.

The third ecozone is located in the tropics from the twenty-fifth parallel north in the west of the Old World to the thirtieth parallel in the east. In China, the Qinling Mountains form a natural barrier to the monsoon and are therefore an approximate marker of the separation between temperate and tropical zones. In the latter, the average annual temperature is always between 70°F and 86°F (21°C and 30°C), and never falls below 64°F (18°C). There is heavy rainfall in summer during the monsoon—i.e., the wet season. Winters are dry. The key point here is that humans from Africa can adapt primarily and most effortlessly to tropical ecosystems. Denisova's ancestors were no exception.

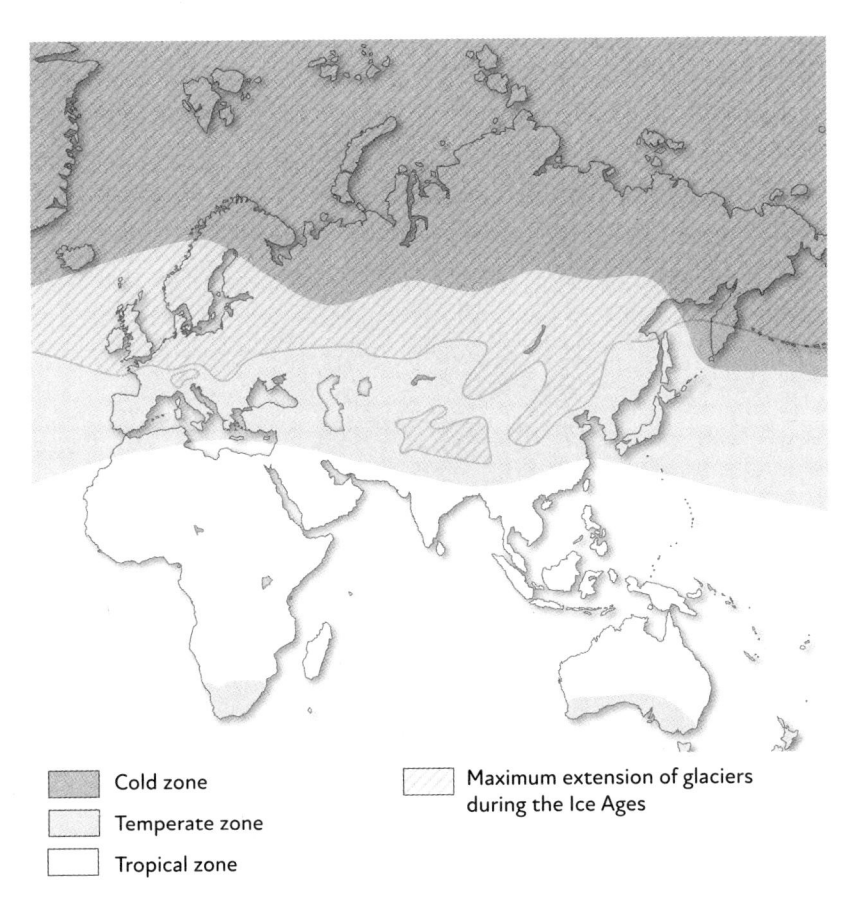

Cold zone

Temperate zone

Tropical zone

Maximum extension of glaciers
during the Ice Ages

Map of the three main terrestrial climatic ecozones in Eurasia and Africa

This reproduction of a young *Homo ergaster* is based on the
most complete skeleton of the species available. It is 1.6 million years old
and contains 108 bones.

8

The Melting Pot That
Made Denisova

The DNA found in the sediments of the Denisova Cave reveals that the first Denisovans ventured into the cold Altai valleys over two hundred thousand years ago, probably much earlier. Their adventure is reminiscent of that of Europeans on the vast North American continent during the conquest of the Americas. The Europeans believed themselves to be far superior to Native Americans, but despite their enormous technical advantages, it took them no less than three centuries to conquer the North American continent. During these three hundred years, pioneers of both French and Native American descent had a crucial role to play.

The French of New France (starting with Samuel de Champlain, 1567–1635, founder of Quebec City) were worried about their small numbers in the face of the "Boston English" and developed an original strategy: They had the Indigenous tribes adopt young French boys. These boys would effectively become bicultural, and serve as intermediaries for Franco-Indian alliances. After the English conquest of Quebec in 1759, the French's open attitude to Indigenous culture continued with the tradition of the *coureurs des bois* ("runner of the woods"). These bicultural French figures spoke several native languages and mastered Indigenous techniques, starting with canoe travel on the Great Lakes and the immense North American river system. They were the first Europeans to be able to explore the North American continent in depth, resulting in thousands of toponyms from French (Detroit, Des Moines, Baton Rouge, St. Louis, Champaign, etc.) or adapted from French (Illinois, Louisiana, Maine, etc.) peppered across Canada and the US.

The French's greater willingness to mix with Indigenous people is reflected by the fact that many Native Americans in the US have French surnames even today. On the Great Plains, there is even a culture that has inherited both European and Indigenous traditions: the Métis people. They are recognized by the Canadian government as a full-fledged Indigenous people, alongside the Inuit and other First Nations people. Today, 53 to 78 percent of Canadians are thought to have at least one Indigenous ancestor. This historical case illustrates how a human wave can spread across a territory: New arrivals follow mixed-race and bicultural pioneers, and then, once there is a large number of newcomers, they assert themselves after absorbing some of the genes and cultural traits of the population that preceded them. The existence of Denny in Denisova—the first-generation hybrid hominin from chapter 2—is an example of the same mechanism and shows that, like the Neanderthals who came to meet them, Denisovans were the children of migrants. But where did their ancestors come from? *Homo erectus*?

Human Evolution: A Constant Melting Pot

The answer is no. But we have one clue as to their true origins: Like the Canadian trappers advancing into the uncharted spaces of the North American continent, the ancestors of the Denisovans mixed with the locals—the "super-archaics." This irrepressible tendency of the *Homo* genus to interbreed is particularly evident in the case of Sapiens. In Paleolithic, Neolithic, protohistoric, and historical times, these humans mixed with all those they found along the way. Each time, Sapiens began by adopting the techniques of their hosts by integrating into their groups, before eventually absorbing these hosts once they outnumbered them. This is the same process that occurred during the European expansions into the Americas, Australia, northern Asia, etc. Far from being a marginal phenomenon, the ability to mix is a crucial social trait of all humans, enabling them to adapt everywhere.

Alan Rogers's team at the University of Utah has clearly demonstrated that interbreeding frequently took place during the Paleolithic period. These paleogeneticists have statistically modeled the succession of generations since the earliest exodus from Africa, using French, English, West African, European Neanderthal, Altai Neanderthal, and Denisovan genomes. The evolutionary model obtained suggests that contemporary human genomes can best be explained if we accept the hypothesis that there were three major successive interbreeding events in Eurasia.

The first occurred between the Denisovans and an unknown population (the "super-archaics" from chapter 4) who are assumed to correspond to *Homo erectus*; the second occurred between the ancestors of *Homo sapiens* and those of *Homo neanderthalensis* more than two hundred thousand years ago, probably in the Levant; the third occurred between late Neanderthals and *Homo sapiens* who had just arrived in Eurasia over fifty thousand years ago. Along with these three instances of interbreeding, there was also the hybridization of Denisovans and Neanderthals in northern Asia. Genetics has in fact confirmed that humans intermingled.

What are the implications of these encounters between different types of humans? Firstly, the arrival of a generalist human form—and all the biology, demography, and material culture it brought with it—was always a slow process. Initially, the encounter between two groups produces bicultural hybrids. These individuals then reproduce, sometimes to the point of creating intermediate cultures of their own. Finally, the vitality of the immigrant population leads to the widespread propagation of its biological and cultural traits in the hybrid population. The original population, itself descended from previous migratory waves, therefore becomes part of the biology and culture of the newer generalist form, which generally become the majority, sometimes long after arriving. Many of the original genes and cultural traits advantageous to survival are retained (positive selection) while those that hinder reproduction are eliminated (negative selection). The spread of a generalist human wave thus has one ubiquitous feature: It is above all one big mixing process, whereby the biological and cultural traits of the demographically dominant incoming population gradually take over from those of the local population resulting from previous waves. This occurs without all traits of the original populations disappearing completely. This process has been repeated several times over the course of human evolution. The arrival of Europeans in the Americas provides a clear example, because it's so recent. The interbreeding of *Homo sapiens* with Neanderthals throughout prehistory is another, which explains why Eurasians possess between 1 and 3 percent Neanderthal DNA. This Neanderthal heritage facilitated the survival of the first Sapiens in Eurasia because it gave them better resistance to the cold and promoted efficient fat metabolism, to give one example. Similarly, Tibetans have retained Denisovan genes that reduce altitude sickness. Various adaptations seen in today's humans also originate from interbreeding with ancient forms.

On the Trail of Generalist Humans

How can we track a wave of generalist humans into Asia? Is genetics the answer? When it comes to the penultimate generalist wave of humans, this is only partially possible because the final wave—that of Sapiens—absorbed its predecessor. That leaves us with nothing but the clues left by groups of hunter-gatherers passing through.

First and foremost are fossils—the most revealing vestiges, but also the rarest. In tropical latitudes, it takes extraordinary circumstances for fossilization to occur, as heat causes rapid and often total deterioration. The case of the main Denisova site illustrates this particularly well. In southern Asia, prehistorians have only two putative Denisovan fossils at their disposal, along with a single tooth.

These are accompanied by a collection of stone tools, which have been perfectly preserved and can be found all across the region. The only practicable way of tracking a human wave out of Africa is to identify and date the types of tools used along the way. This method has worked very well in Africa, Europe, and India, but it ceases to be effective farther afield in Asia. This is because the adoption of techniques brought by migrants is slow, and these time lags need to be taken into account. In Asia, the part of Eurasia farthest from Africa, these lags become very significant, and as we'll see, tracking lithic tools proves very difficult past the eastern gates of India. Beyond that, the trail gets lost. How to overcome this issue? The best method is to analyze the traces that have been left behind, bearing in mind that within the *Homo* genus, three evolutionary laws still apply:

- Cultural activities are what distinguish human lineages from animal lineages.
- The African origin of human lineages suggests that the evolution from one cultural stage to another in Africa precedes the same evolutionary stage in Eurasia.
- The arrival of a new kind of biology precedes cultural renewal, often by a long time, as the latter spreads from Africa or is induced by the local environment.

From Biological to Cultural

Now we've identified the biology of at least four generalist human waves (the African forms of *Homo habilis, Homo erectus,*

Homo rhodesiensis/heidelbergensis, and *Homo sapiens*), we must also define four successive cultural stages associated with them. For research like this on a global scale, our preferred method is the simplistic but effective system of British prehistorian Grahame Clarke (1907–1995). This scholar was one of the first to organize the chaos of the multiple types of lithic industries, in order to reveal major cultural evolutionary stages. These are clearly defined in Europe and certain regions of Africa but not necessarily everywhere else. Clarke differentiates between four major successive stages in tool-cutting practices (see the four main tool types, pages 127–29):

- Mode 1 is linked to the lithic industries known as Oldowan (2.8 million years ago in Africa), used by the first human artisans in Africa: *Homo habilis*.

- Mode 2 is linked to Acheulean industries (1.7 million years ago in Africa), used by *Homo erectus* in Africa.

- Mode 3 is linked to the production of tools made from lithic flakes, used by *Homo rhodesiensis/heidelbergensis* and their descendants.

- Mode 4 is characterized by the evolution of flaked blades toward the microlith, used by *Homo sapiens* alone.

Although Clarke's system is not universally accepted and there is some chronological overlap, it's very useful for understanding human cultural evolution in archaeological data. And while the modes are effective for analyzing prehistoric human behaviors in Africa (except in the forest), throughout Europe, India, and central and southern Asia, the same does not apply in tropical Asia.

The First Generalist Wave?

For the first wave to have swept across Eurasia, it must have done so in Africa beforehand. The most obvious candidate in the running to be named the first generalist human is *Homo habilis*, which existed 2.3 to 1.5 million years ago. This biped, measuring 4.3 to 4.6 feet (1.3 to 1.4 m) and weighing 77 to 99 pounds (35 to 45 kg), had a cranial capacity of only 550 to 700 cubic centimeters. Was this the first human form to come out of Africa?

Evidence seems to suggest so: A fossil form comparable to *Homo habilis* passed through Dmanisi, Georgia, around 1.8 million years ago. There is nothing in Europe to indicate that these humans traveled westward. If anything, they seem to have moved eastward, though the fossils old enough to plausibly be part of the first generalist human wave in Asia are difficult to link to *Homo habilis*. In Longgudong, for example, fossilized teeth have been discovered that most likely belonged to an archaic human—possibly *Homo habilis*. Another, even more significant discovery is that of the oldest human skull in China: the "Lantian Man." Highly deformed but nevertheless evidently archaic, the skull was unearthed at the northern limit of the monsoon region in Gongwangling, Shaanxi. Its discovery alongside Oldowan tools led researchers to estimate its age at between 1.6 and 1.54 million years old. While fossils are hard to come by, surprisingly old Mode 1 (Oldowan) tools have been found in continental Asia. In China's Chongqing province, Longgupo, a site has been found with Oldowan (single-sided flint) tools that, astonishingly, date back some 2.2 to 2.5 million years. The Renzidong site has yielded some equally astounding discoveries, dating back some 2.25 million years. The presence of humans in eastern Asia over 2 million years ago has also been confirmed by two sites featuring single-sided flints: Shangchen in Shaanxi (2.12 to 1.26 million years old) and Longgudong (2 million years old) in Hubei (see the map of the first migration out of Africa on the following page).

While it's likely there was such a thing as a generalist first wave, this remains to be clarified, as the dates are so uncertain. What's more, the first wave's departure from Africa could date back to a time when the distinction between humans and Australopithecus was tenuous.

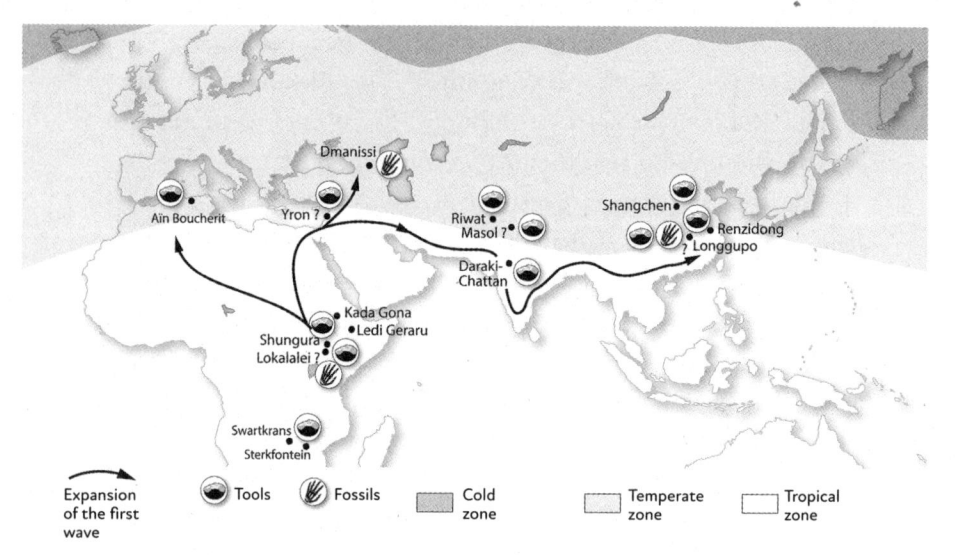

Map of the first dispersal wave of the *Homo* genus out of Africa (over 2 to 1.5 million years ago): The first generalist human form followed the southern shores of Eurasia and also took advantage of a warm period to venture north, eventually reaching the Georgian site of Dmanisi.

A Known Second Wave

What direction did *Homo erectus* take when it left Africa? For a start, at the Ubeidiya site in Israel, a vertebra of an adolescent, two teeth, and some hand axes confirm that members of the second wave of generalist humans using Mode 2 (Acheulean) tools lived in the Levant 1.5 million years ago.

The 1.3-million-year-old skull from Kocabaş, found in isolation and without other contextual clues in Turkey, suggests a northward progression of *Homo ergaster*/*erectus*—who were using Mode 1 (Oldowan) tools—toward Europe, a continent only 1,200 miles (2,000 km) from the Levant.

This wave is the first example of a clear lag between traces of the arrival of a generalist human form and traces of the lithic mode they were practicing in Africa appearing in Europe. This human form seems to have progressed along the southern coasts of Africa, leaving behind stones and other Mode 1 (Oldowan) tools in a series of sites. The most easterly of these is the Kozarnika Cave in Bulgaria, geomagnetically dated between 1.6 and 1.4 million years old. A *Homo ergaster/erectus* clan also left behind a molar and a series of cut stones. Beyond that, a broken chain of localities links southeastern and southwestern Europe: Pirro Nord in Southern Italy (1.6 to 1.2 million years ago); Lézignan-la-Cèbe, Lunéry Rosières, Pont-de-Lavaud, and Vallonnet in France (1.2 to 1 million years ago); Barranco León, Fuente Nueva 3, and Sima del Elefante in Spain (1.2 to 1.4 million years ago). At Sima del Elefante, north of the Iberian Peninsula, fragments of the mandible and face of the oldest known European have also been found (see the map of the second migration from Africa, page 127).

To the east, meanwhile, it seems that members of the second wave of generalist humans had already progressed to Mode 2 (Acheulean) tools in addition to Mode 1 (Oldowan). Evidence of this can be found, for example, at the Attirampakkam site in the Southern Indian state of Tamil Nadu. Did *Homo erectus*

The Ubeidiya site in Israel's Jordan Valley is atypical in that its geological layers are vertical due to tectonic forces, so excavators work standing upright.

spread farther east via the tropics? The large number of Acheulean sites in Eastern India over 1.4 million years old strongly suggests so. What about farther afield? The sites in Java we have already mentioned attest to the arrival of the second wave of generalist humans at least 1.3 million years ago, if not earlier. In Indonesia and the Philippines, prehistorians have collected large numbers of hand axes from the ground's surface. Gustav von Koenigswald reported having found them "by the crateload" at the Sangiran site. Most prehistorians today, however, refuse to date these too far back in time, claiming that these artifacts were flaked by *Homo sapiens* just 12,000 years ago. . . . Even if this were to be proven true one day, it doesn't rule out the possibility that *Homo erectus* continued its trajectory eastward.

Excavations at Wolo Sege, an open-air site located in the basal deposits of the Ola Bula Formation in central Flores, Indonesia, have yielded three hand axes dating back over a million years: Acheulean-type tools made from large lithic flakes. Alfred Pawlik and his colleagues from the University of the Philippines have also excavated the Arubo site on the island of Luzon, and found authentic Acheulean hand axes that, in Africa, would date back to the start of the Paleolithic Age. In short, there appear to be hand axes flaked from lithic cores in Southeast Asia, but the hypothesis that they are over a million years old remains disputed.

This can probably be explained by the fact that in the rainforest-covered lands of Southeast Asia, high-quality stone deposits were difficult to come by. Meanwhile wood—and especially bamboo—opened up the possibility of creating "lignic" tools rather than lithic. We'll come back to this point in greater detail when we visit the third wave of generalist humans: those who managed to penetrate the Asian rainforest.

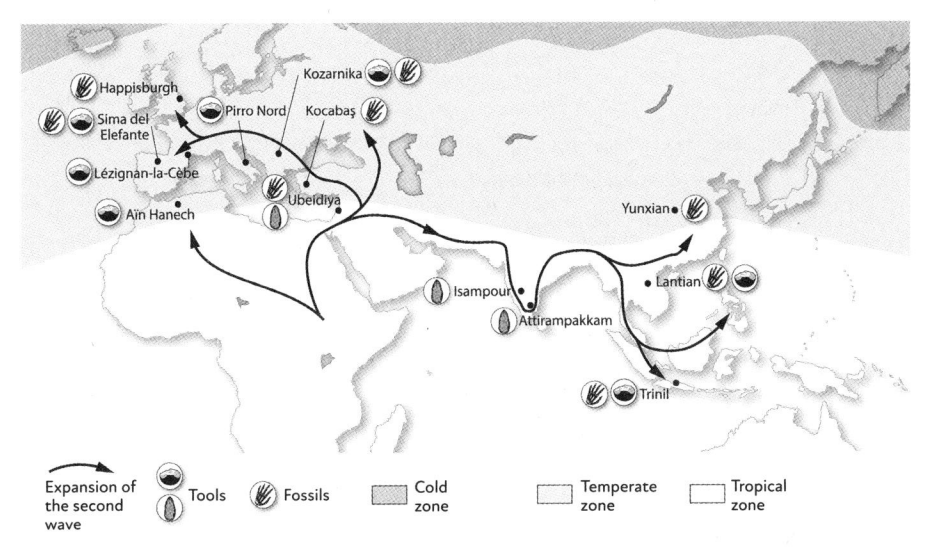

Map of the second dispersal wave from Africa of the *Homo* genus
(1.5 million to 800,000 years ago): Like the first form before it, the second
generalist human form advanced toward the southern shores of Eurasia.
Members of this wave were the first humans in Europe.

Four Major Tool Industries

The diversity of prehistoric lithic industries makes them difficult to
navigate. In order of age, Clarke's "lithic modes" are as follows:

- Mode 1 is the oldest. In Africa, Oldowan lithic industries
  appeared as early as 2.8 million years ago, probably among
  prehumans. Within the *Homo* genus, they were first used by
  Homo habilis (preceded by tool cutting *Australopithecus*).
  In this type of cutting, pebbles are shaped on one side only
  to produce a cutting edge (chopping tools) and sharp lithic
  flakes. Evidence suggests that these tools were primarily used
  to break the long bones of herbivores to recover the marrow,
  or to scrape off the residual flesh left by large carnivores. They
  therefore appear to have been primarily the tools of human
  scavengers.

- Mode 2 is linked to Acheulean lithic industries. *Homo ergaster/erectus* was the first to use these as early as 2 million years ago, followed by *Homo rhodesiensis/heidelbergensis* 800,000 years ago, who perfected the form. This lithic mode features stones that are usually shaped on two sides, sometimes producing highly symmetrical hand axes or choppers. Debitage (carving of lithic cores) also produces large, sharp flakes, which, in turn, can be shaped into tools themselves. These "flake tools" were used for cutting, sawing, scraping, etc., but could also be used in the same way as hand axes, as multifunctional tools for hunter-gatherers, who were no longer just scavengers.

- Mode 3 features the debitage of lithic flakes, and is thought to have first been used in Africa by *Homo rhodesiensis/heidelbergensis*. This mode is associated with the Middle Stone Age of eastern and southern Africa (Middle African Paleolithic or Middle Stone Age, between around 400,000 and 50,000 years ago), as well as the Middle Paleolithic of Europe, North Africa, and western Asia (around 350,000 to 45,000 years ago). Mode 3 is also known as the Mousterian industry, featuring techniques shared by Sapiens in North Africa and the Levant, and Neanderthals and Denisovans in Eurasia. This mode involves carving flakes from a lithic core, some of which would be subsequently refined, sometimes using antlers as a form of soft hammer, to form "knives" with sharp blades, scrapers, or hide scrapers. While the occasional hand axe has been found on sites associated with this industry, Mousterian tools largely consist of blades that were used as knives or components for projectile weapons—i.e., composite tools. These were sometimes fashioned in part using fire. Humans who had advanced to Mode 3 lithic industries were hunters.

- Mode 4 is characterized by an evolution of blade production toward microlithism and was adopted solely by *Homo sapiens*. It is associated with the Late Stone Age (LSA, from around 50,000 to 25,000 years ago) in sub-Saharan Africa and the Upper

Paleolithic in Europe and Asia (50,000 to 25,000 years ago). Mode 4 industries are also known as Aurignacian. It primarily consisted of the production of blades used either as knives or as armatures for projectile weapons and was also characterized by the heavy use of bone. Mode 4 and Mode 5 are, as far as we know, associated solely with Sapiens and therefore of little pertinence to our investigation following in the footsteps of Denisova.

The China Enigma

What about continental Asia? The problem is that it's difficult to date sites in this region, and any attempts to do so date back to when methods were not very advanced. While we have associated the skull of the "Lantian Man" with the second wave of generalist humans, we cannot be sure, given its age of 1.6 million years and its severely deformed state. Aside from this first fossil, the oldest remains found in China come from Xuetang Liangzi, Hubei. The site is in Yunyang district, around half a mile (100 meters) from the Yan River. In 1989, Wang Zhenhua, curator of the Wuhan Archaeological Museum, came to excavate the site after a local farmer discovered a fragment of elephant tusk there. The first skull, *Yunxian 1*, was dug out of very hard sedimentary rock, littered with flints cut on only one side (a handful were cut on two sides). Since then, Chinese prehistorians have unearthed over 2,000 bones from more than 20 rainforest species, including a giant panda, an elephant and others. Following these discoveries, Li Tianyan's team from the Hubei Institute of Cultural Relics and Archaeology systematically excavated the site and found a second skull, *Yunxian 2*, 10 feet (3 m) from the first, in the same stratum, only 40 centimeters deeper. Finally, in 2022, just over 100 feet (35 m) from the other two fossils, a third skull known as

Yunxian 3, was discovered in a similar context. From the enamel of animal teeth found in the same stratum as *Yunxian 2*, dates for the fossils were determined using highly reliable paleomagnetic and electron spin resonance techniques. These indicate that the fossils date back 936,000 years.

Unfortunately, skulls 1 and 2 are highly deformed by the weight of the sediment, but 3D computer models have been able to provide us with information on their general shape. Unsurprisingly, the skulls are low and elongated, as in all archaic humans. The forehead is heavily receding, while the maximum width of the skull is narrow, producing an almost triangular cranial vault. Cranial capacity is estimated at around 1,050 cubic centimeters, clearly placing the Yunxian humans on the *Homo erectus* spectrum. The mixture of archaic and derived traits in Yunxian humans indicates they evolved locally, after the arrival of the second wave of generalist humans in Asia. Nevertheless, these are still a form of *Homo erectus*.

In addition to the Yunxian skulls, China has produced a number of Oldowan industry sites. On these sites, human fossils have either not been found, been too fragmentary, or badly dated. These do not belong to the Denisovan chronological bracket, which begins around four hundred thousand years ago (Yuanmou, Guojiabao, Coloquinte, Chenjiawo, Yiyuan, etc.). The striking thing about these sites is that they are primarily situated to the north of China's tropical climate zone. They were found in the places where chances of preservation were highest, away from the warm humidity of areas affected by monsoon, as fossilization is a rare phenomenon in a truly tropical climate. This observation supports our idea that second-wave generalist humans first followed the southern shores of Eurasia, spread into Southeast Asia, then moved northward into present-day China. The locations of sites they once inhabited suggests they tended to remain in fairly warm climates, as they were maladapted to the cold.

Was the Peking Man also a member of the second wave of generalist humans? The oldest specimens of its kind indicate cranial capacities typical of the late second wave of generalist humans—less than 1,100 cubic centimeters. This would make them around 780,000 years old, suggesting these visibly archaic humans were *Homo erectus*. Their presence near Beijing, in a temperate continental climate with cold winters, implies that the last members of the second wave of generalist humans had evolved sufficiently in biological terms to enable them to withstand the cold weather using clever techniques (see the text box that follows). Perhaps, as the excavators of layer 10 at Zhoukoudian in the early 20th century claimed, they even had fire. Whatever the case, their successors—Denisova's ancestors—had completely mastered it.

Fat: An Effective Means of Keeping Warm

The people of Tierra del Fuego Darwin encountered in 1832 demonstrate how it might have been possible for members of the second wave who had ventured north to live in cold conditions. Of course, those Sapiens living in Tierra del Fuego in 1832 had fire. But these hunter-gatherers would often fish naked despite the icy cold conditions of their region. They rarely wore clothing, instead protecting themselves from the cold, and worse, the damp, by coating their bodies in animal fat. Soviet archaeologists have proven that prehistoric people had mastered techniques for withstanding low temperatures by dating their arrival to the plains and mountains of Kazakhstan, which are freezing in the winter, to between nine hundred thousand and six hundred thousand years ago.

In fact, as early as 1966, Stanislav Arkhipov of the Russian Academy of Sciences reported that the first signs of human presence along the Irtysh River date back some 0.8 million years. In the "Black Mountains" of Karatau, a Lower Paleolithic culture that had adopted Mode 1 (Oldowan) industries, may even have existed around a million years ago. These dates correspond with findings from Happisburgh, England, where human footprints between 1 and 0.8 million years old were found fossilized on a beach. The footprints were unearthed by a storm of exceptional intensity and were photographed before being permanently erased by the tide.

What this tells us is that a small fraction of the second wave of generalist humans found their way to the steppes of the temperate belt spanning Europe and Asia, which were teeming with herbivores, probably by traveling both from west to east and east to west. In fact, the hunter-gatherers of Zhoukoudian had already inhabited the cold Beijing plain some 780,000 years ago. In the 1920s, the first excavators of the site in Zhoukoudian believed they had found evidence of pyrotechnics in the site's oldest strata, but this is now considered insufficient. What if they were right?

European *Homo heidelbergensis* were the ancestors
of *Homo neanderthalensis*.

9

Denisova and Neanderthal: A Common Ancestor

The site at the Gesher Benot Ya'akov ("Daughters of Jacob Bridge") on the Jordan River is known to prehistorians by its three initials: GBY. This 790,000-year-old, open-air habitat has been known since the 1930s and is of major scientific significance. In 2009, Naama Goren-Inbar's team at the Hebrew University of Jerusalem revealed that the ancient inhabitants of the site divided their living space into specific zones for cooking, carving tools, etc. According to the researchers, these behaviors imply advanced cognitive abilities linked to larger brains.

This advanced level of cognition is particularly evident in the shaping of numerous stone tools of a more "evolved" Acheulean size. In other words, the hand axes that have been discovered at GBY are highly symmetrical—much more so than those of Ubeidiya, a site associated with the second migratory wave from Africa. These tools are evidence of the arrival of new humans. The traces they left on the site are evidence of a culinary innovation: The inhabitants of GBY had mastered the art of cooking fish.

The team of researchers at GBY found that the site's hunter-gatherers cooked giant carp of up to 6.5 feet (2 m) long, which they caught in a nearby lake. Several burnt flint microartifacts suggest they made fires. In the same layers as the burnt flints, the researchers collected around 40,000 pharyngeal teeth, carp teeth that are found at the bottom of their mouths. These teeth are all that remain of the carp, as heat softens the cartilaginous bones of fish and eradicates the possibility of their preservation. The discovery of these teeth suggests the fish were cooked at a controlled temperature rather than being grilled.

Even if we don't know how the occupants of GBY made their fires, we do know that they stewed fish. Using X-ray diffraction, researchers established that the thermal expansion of the nanocrystals that make up the tooth enamel suggests their exposure to low to moderate heat: specifically, lower than 932°F (500°C), whereas a wood fire produces temperatures of between 1,472°F and 1,832°F (800°C and 1,000°C). It's likely that prehistoric

View of the excavations at the Gesher Benot Ya'akov ("Daughters of Jacob Bridge") archaeological site in Israel, known as GBY

people cooked carp *en papillote*, probably by burying them near their fires after wrapping them in giant water lily leaves collected from the nearby lake.

The superior cognitive faculties implied by this cooking method, the mastery of fire, and the shaping of advanced Acheulean tools show that the GBY humans, fresh from Africa, are different from those of the second wave of generalist humans. The inhabitants of GBY were new humans who entered Eurasia around 800,000 years ago, a fact confirmed by geneticists, who believe that the split between the African and Eurasian branches occurred around 750,000 years ago. After that, the third generalist human form would continue to evolve, giving rise to specialist forms: Sapiens in Africa, Neanderthal in Europe, and Denisova in Asia.

We've already met these humans in chapter 4: *Homo rhodesiensis/heidelbergensis*. Once they arrived in Europe, *Homo rhodesiensis* is referred to as *Homo heidelbergensis*, even though they are the same species.

Homo heidelbergensis in Africa

Let's start from the beginning. In the cradle of humanity, fossils associated with the third generalist form of humans range from seven hundred thousand to three hundred thousand years old. British prehistorian Arthur Smith Woodward named the species *Homo rhodesiensis*, taking the name of a fossil known as "Rhodesian Man." In 1921, Swiss miner Tom Zwiglaar and an African colleague whose name history has unjustly forgotten discovered the fossils in the Broken Hill mine, now known as Kabwe, in Rhodesia, modern-day Zambia. Among them were a skull, a maxilla, a sacrum, a tibia, and two femur fragments. The destruction of the site by mining activities prevented any stratigraphic dating, so the age of the fossils was regularly debated during the twentieth century, until direct dating of the skull established an age of around three hundred thousand years old.

Given the date Sapiens appeared, this comes as a surprise; it means the last of the *Homo rhodesiensis* were still alive and well in Africa when *Homo sapiens* emerged. This once again demonstrates that evolution is more akin to a tree with many branches than a line, as *Homo sapiens* both evolved from *Homo rhodesiensis* and rubbed shoulders with some of its last remaining members.

This might seem confusing at first, but fortunately, the third wave of generalist humans has also been identified by older fossils scattered across Africa. A few examples include the Bodo skull found in Awash, Ethiopia (six hundred thousand years old) which is very similar to the skull found in the Kabwe mine; a third found in Saldanha in southern Africa (six hundred thousand years old); and a fourth specimen found at Lake Ndutu in Northern Tanzania (five hundred thousand years old). The other African fossils traceable to *Homo rhodesiensis* are two mandibles, a parietal bone, and teeth, estimated to be seven hundred thousand years old, from Tighennif, exhumed in the 1950s near Oran, Algeria.

What kind of a portrait can we sketch of *Homo rhodesiensis*, given this information? The fossils, when put together, suggest a tall, slender form, which means these humans were adapted to tropical conditions. At between 1,100 and 1,300 cubic centimeters, its cranial capacity is within the range of variability for Sapiens. There are both archaic and evolved forms of *Homo rhodesiensis*. Its archaic features include a very large face with prominent supraorbital reliefs and a narrow, receding frontal bone, bulging, elongated parietals—the bones at the top of the skull—and very pronounced bone reliefs throughout the cranial vault. The most striking evolutionary feature, which brings *Homo rhodesiensis* closer to its descendant *Homo sapiens*, is its large cranial capacity. This is where we arrive at what paleoanthropologists describe as the "muddle in the middle." This curious expression is used in reference to Old World fossils from the period between 800,000 and 100,000 years ago, whose features—sometimes similar and sometimes distinct—have caused much confusion.

Despite their small number, the age, and wide distribution of *Homo rhodesiensis* fossils throughout Africa demonstrate that, by 800,000 years ago, this human had settled across most of Africa. In other words, it had become a generalist human. Once it had become adapted to all the many African climates and ecosystems, it was ready to move into Eurasia, which it did starting from 800,000 years ago.

The Hunt for Hand Axes

How to track a human form across this immense continent? By a method you may be familiar with by now: hunting down fossils and cultural markers. As we've seen in the case of GBY, the third wave of generalist humans is characterized not only by a new kind of biology identified from fossils and evidence of the use of fire, but also an evolution in tools to hand axes (double-sided blades).

We've already mentioned hand axes, prehistory's answer to the Swiss Army knife, and touched upon how the expansion of the second wave led to their spread outside of Africa to the Near East and India, probably as early as 1.5 million years ago. These hand axes were asymmetrical and very voluminous, unlike those from GBY, which marks the beginning of an evolution toward more and more tools with bifacial and bilateral symmetry. The best examples can be found in Europe, with flint hand axes of a spectacular quality dating back some 500,000 years (see the hand axe on page 6 of the insert).

The second salient feature of the lithic industry of the third wave is the production of very sharp blades. These "knives" are no longer just simple lithic flakes resulting from knapping, but the deliberate result of an optimized chain of operations, reflecting a cognitive complexity requiring mental previsualization of gestures and a production strategy. Blade production was perfected from five hundred thousand years ago in eastern and southern Africa, eventually reaching the high efficiency of lithic flake debitage practiced in Europe. This technique involves knapping a series of flakes from the lithic core, thus obtaining the greatest possible length of cutting edge from each stone.

Fire: The Father of Ingenuity

In addition to tools, the second cultural marker of the third wave of generalist humans is the mastery of fire, an innovation that has had major effects on human evolution, firstly by providing access to non-tropical climates. After the third wave left Africa 800,000 years ago, there are only a few traces of fire. GBY is an exception in this regard. The remains of fires that have been found are perhaps examples of "tamed" fire (i.e., fire that humans took advantage of without necessarily understanding or knowing how to make it), which had to be constantly maintained. Then, from 400,000 years ago onward, traces of controlled fires became ubiquitous, which demonstrates mastery of fire. By this point, humans knew how to produce it at will and use it extensively for technical purposes. We don't know when the ability to light fires was acquired, but the consensus is that the oldest known controlled fires can be found at GBY, dating back 790,000 years.

What about Europe? GBY, where fires were in use for some fifty thousand years, suggests that *Homo rhodesiensis/heidelbergensis* was already making daily use of fire before migrating to the temperate zones of Eurasia. They likely tamed fire and perhaps even learned to master it.

Whatever the case, fire was present in Europe as early as 700,000 years ago at Přezletice, in Czechia, then around 450,000 years ago at Vertessöllös, in Hungary, and at Menez Dregan, in Brittany. From 400,000 years ago, it seems to have been in constant use. If, between 700,000 and 400,000 years ago, *Homo heidelbergensis* was already present in Europe, it was probably still poorly adapted to the cold, so it would have moved to the temperate band during warm periods and retreated southward during cold periods. The period of time between GBY and the appearance of fire in Europe is therefore probably equivalent to the time it would have taken to learn to cope with the cold. Increasing mastery of fire most likely played a crucial role in this process. In Asia, there are traces of fire at the Zhoukoudian site around 400,000 years ago, and likely even earlier. The same goes for many other sites.

The mastery of fire brought great technical advantages, as illustrated by two European sites in particular: Clacton-on-Sea in England, 400,000 years old, where hunters are known to have sharpened a spear before hardening its tip with fire, and the 300,000-year-old Schöningen site in Germany, where *Homo heidelbergensis* hunters left eight pine or spruce spears on a lakeshore. Ranging from 5.9 to 8.2 feet (1.8 to 2.5 m) in length, with a maximum diameter of 2.9 to 4.7 centimeters, these spears were found close to horse skeletons, which were undoubtedly the hunted prey. The study of these fire-hardened weapons reveals that they were carefully flaked to make the point solid for spearing prey in flight, suggesting that the hunters applied ballistic knowledge.

Among the European descendants of *Homo heidelbergensis*— the Neanderthals—mastery of fire is even more evident. In 2018, a team led by Andrew Sorensen from Leiden University in the Netherlands, demonstrated that they employed a kind of "fire kit" that clearly facilitated intentional creation of fire. At several Neanderthal sites, these researchers found macroscopic and microscopic traces on dozens of hand axes, suggesting vigorous percussion and abrasion using hard mineral materials. Experimentation suggests that the occupants of these sites struck the hand axes obliquely with fragments of pyrite (FeS2), so as to project sparks into a flammable material (tinder fungus, no doubt). Once dry, this spongy substance derived from several fungi becomes flammable. We know that Ötzi, a Bronze Age man who died five thousand years ago in an Italian glacier, was carrying some with him. Another technical use of fire: Neanderthals had developed a thermal process to extract a natural tar—birch pitch—from birch bark. They used it to attach blades to the handles of their projectile weapons, and no doubt for many other purposes.

Culture, the Mother of Human Anatomy

All this makes it clear that the mastery of fire and the increased cognitive capacities of third-wave generalist humans are closely linked. Let's go a step further and look at what the ability to cook has changed for human biology. Cooking food is a form of pre-digestion: It increases the caloric value of certain foods. Studies show that humans digest 35 percent of cooked starch, but only 12 percent of raw starch, and 78 percent of cooked proteins, compared with only 45 percent of raw proteins. Cooking also has the advantage of disinfecting and rendering edible many foods that would otherwise be toxic, such as acorns from oaks, many natural legumes, fish that isn't fresh enough, many types of mushrooms, and so on.

Cooking effectively accelerated the process of evolution. In 1996, Leslie Aiello from the University of London, and her colleague Peter Wheeler from the University of Liverpool, studied the constitution of the body in relation to the energy cost of our organs, observing that the size of the intestine correlates with that of the brain. The researchers thus formulated the "expensive tissue hypothesis"—i.e., the idea that the human brain owes its large size to redistribution of energy from other organs in the digestive system. In fact, despite representing only 2 percent of a human body's mass, the brain consumes 20 percent of its energy.

This reorganization has led to the reduction of the digestive system by 40 percent compared to other primates, so that today, digestion requires barely 10 percent of the energy necessary for the basic functioning of our organism. This has also led to a reduction in the size of our chewing muscles, and their attachment points on the face and skull, as well as the positive selection of human lineages that require less time to chew. These adaptations allocated more energy to cognition. In short, there's no need to chew things over, just take a little time to digest. . . .

Aiello and Wheeler's theory has been confirmed and refined by other researchers to give us a clear evolutionary history of the

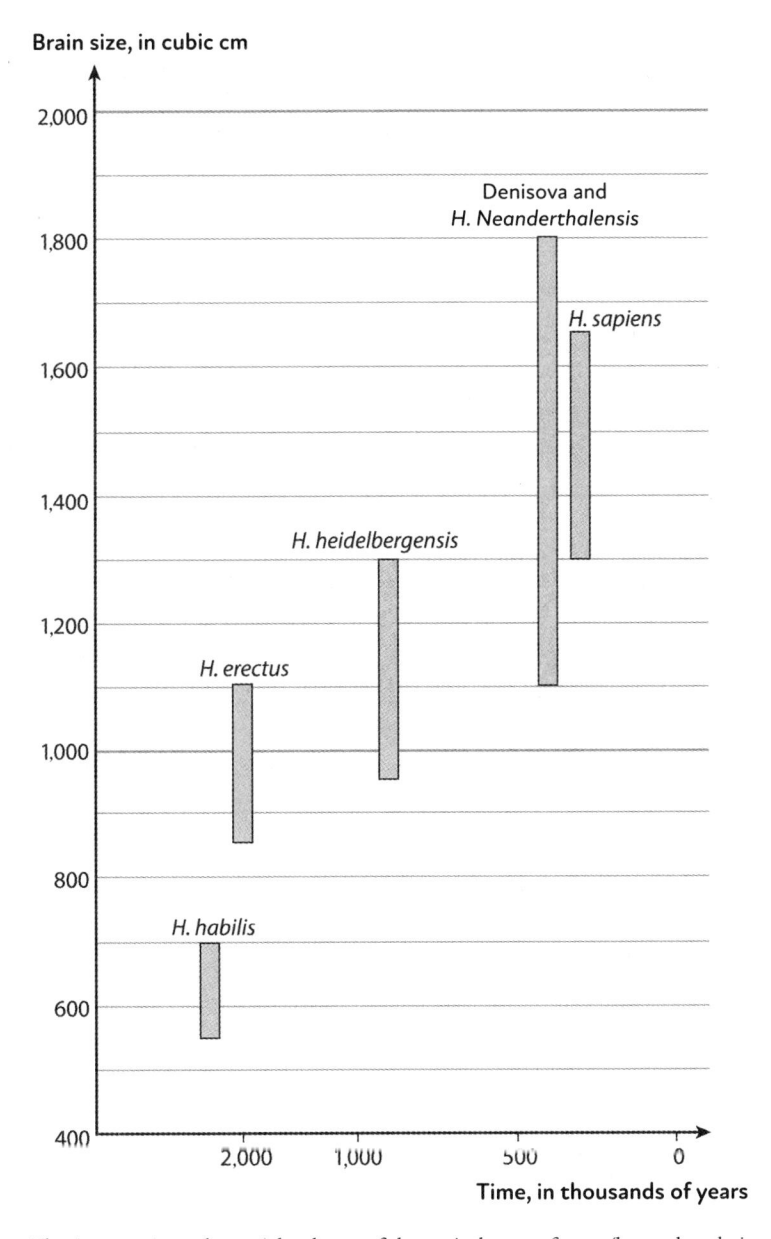

The increase in endocranial volume of the main human forms (located at their estimated date of emergence/estimated date they first appeared) over the last 2 million years reflects the increase in brain size, and, a priori, of cognitive capacities.

human brain today: While our ancestors were eating plants and raw meat, brain volume increased only in proportion to stature. From the moment they began to consume cooked foods, however, cranial capacity ceased to be proportional to this factor. Brain size then continued to grow until it reached volumes of the kind seen among humans today. In *Homo heidelbergensis*, the fundamental metabolic rearrangements that led to the species attaining an optimal height (from 5.2 to 5.9 feet/1.6 to 1.8 m in males) and a large brain capacity (from 1,100 to 1,300 cubic centimeters) were thanks to cooking. Brain size then continued to increase in the descendants of *Homo heidelbergensis*, namely the African Sapiens, the European Neanderthal, and the Asian Denisova: an evolution that occurred thanks to culture.

In the Footsteps of *Homo heidelbergensis*

Fossils, hand axes, and fire can all help us track down *Homo heidelbergensis* in Europe. In fact, its arrival was first signaled by evolved hand axes, then by fossils accompanied by basic hand axes. The oldest known hand axes, which are 700,000 years old, are those from Notarchirico-Venosa in Southern Italy and La Noira in France (evolved hand axes); they are followed by those less than 650,000 years old from Caune de l'Arago and Moulin Quignon in France, Gran Dolina in Spain, and Warren Hill in England. The introduction of debitage techniques also reflects cultural renewal.

As far as fossils are concerned, the oldest is the Mauer mandible, discovered in 1907 by a worker in a sand quarry near Heidelberg, Germany. Its age was estimated in several laborious stages at six hundred thousand years old. It was not known at the time, but this massive mandible with its voluminous teeth, which gave the species *Homo heidelbergensis* its name, was the first evidence of the arrival of the third wave in Europe. It's worth noting in passing that, had *Homo rhodesiensis* been discovered first, its name would likely have been used for the species instead, but as European conditions are far more favorable to fossilization than those in the tropics, Europe provided the best chance of finding biological traces of *Homo heidelbergensis*—and there are plenty of them. In addition to the Mauer mandible, twenty-eight fragmented skeletons have been discovered in Atapuerca, Spain; skulls and other mandibles and bone fragments in Caune de l'Arago, France; a tooth in Vergranne, France; a magnificent skull in Petralona, Greece; a tibia in Boxgrove, England; a skull in Ceprano, Italy; and so on. See the following map for an overview.

Like their African counterparts, European fossils from the third wave of generalist humans have a large cranial capacity. The 500,000-year-old Boxgrove tibia also illustrates that *Homo heidelbergensis* long retained the elongated, heat-expelling limbs of humans in African climates. Its owner was no less than 5.7 feet (1.75 m) tall. This tibia and the powerful Mauer mandible therefore tell us that half a million years ago, the members of the third wave were still biologically close to their African ancestors.

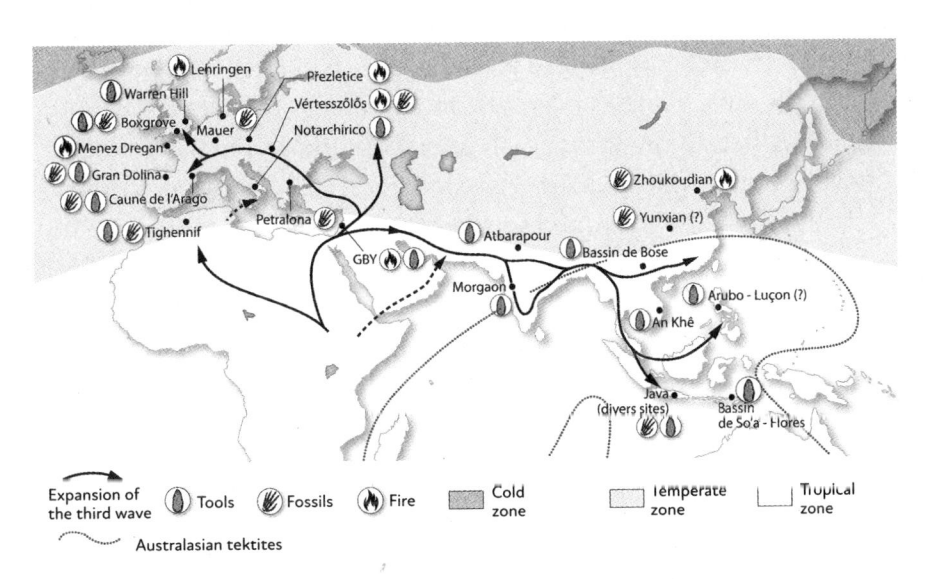

Map of the third dispersal wave of the *Homo* genus from Africa (800,000 to 300,000 years): Like the forms that preceded it, the third generalist human form first advanced to the southern shores of Eurasia, but, having mastered fire, it was also able to set up camp in the temperate zone.

In the three hundred thousand years following the arrival of *Homo rhodesiensis/heidelbergensis* in Eurasia, however, this human form must have interbred with the *Homo erectus* that preceded it. In 2015, Portuguese prehistorian João Zilhão's team discovered the four-hundred-thousand-year-old skull of Aroeira, with a "mosaic" of features from both *Homo erectus* and *Homo heidelbergensis*. The case of "Portugal's oldest man" seems to provide evidence of interbreeding between successive generalist human forms.

A Clear Parallel Evolution

Why is Europe so prevalent in our investigation into the emergence of Denisova? For one simple reason: Because Neanderthal is clearly close to its Asian sibling; all genetic studies confirm that the two evolved at the same time. As they did so, each in its own part of Eurasia, they only began to grow distinct from one another after four hundred thousand years. The major geographical differences between Europe and Asia explain the difference between the two related populations of Neanderthals and Denisovans.

In Europe, pre-Neanderthals found themselves confined by the Mediterranean during each glacial maximum and had to survive in very cold climates. As we have seen in relation to the Earth system, by the time *Homo rhodesiensis/heidelbergensis* arrived in Eurasia, the rate of glaciation had increased from forty thousand to one hundred thousand years, resulting in the north of the continent being frozen over for much longer periods. Under these conditions, it was not possible for large human populations to live at high latitudes.

In addition, the glaciations of marine isotope stages 12 (478,000 to 424,000 years ago) and 10 (374,000 to 337,000 years ago) were extremely rigorous and long—54,000 years, or 2,700 generations, and 37,000 years, or 1,850 generations, respectively (5 generations per century). In the west, pre-Neanderthals were trapped by the Mediterranean in several genetic bottlenecks, while in the east, pre-Denisovans always had the option of crossing the Qinling Mountains back into tropical conditions. The lack of diversity in the Neanderthal genome in comparison to Denisovans is a reflection of these episodes of extreme survival in Europe. This difference between the sparse Neanderthal population and the Denisovan population, whose survival was secured by a vast reservoir in the tropics, has not gone unnoticed among geneticists. Denisova and Neanderthal are related in that they are both descended from *Homo rhodesiensis/heidelbergensis*. But the fact they both evolved in parallel, each in its own corner of Eurasia in very different ecosystems, no doubt explains much of what distinguishes them.

From 350,000 years ago onward, human remains appear that feature Neanderthal traits, so we can expect to find fossils with Denisovan traits from around the same date in northern Asia. We know that Neanderthals survived extreme ice ages. The ways in which they evolved to adapt to the cold can be seen in their stocky bodies with short arms and legs: A massive torso conserves internal heat better, while smaller appendages dissipate it less. During the 300,000 years of their evolution, Neanderthals remained few in number, but they were so well adapted to the harsh European conditions that they managed to endure the three severe ice ages of the period and survive until around 40,000 years ago. Certain Denisovans also adapted to the cold as attested by the dates of DNA traces found in the Denisova caves in the Altai and the Baishiya caves on the Tibetan plateau. It is clear, however, that most of the Denisovan population thrived in the tropics and periodically came into contact with their relatives from the north when they ventured farther afield during warmer periods. While Neanderthals found themselves confined to a small, freezing territory during glacial maximums, Denisovans continued to thrive across an immense continent that had expanded due to decreasing sea levels, and still had enough exchanges with their northern relatives to maintain their genetic diversity. This difference between western and eastern Eurasia helps us to understand how Neanderthal and Denisova evolved in parallel: The

Denisovans that had settled in northern Asia were protected from genetic bottlenecks, even if their living environment was only a fairly marginal zone of the immense territorial range their species inhabited. This territory was also much more conducive to fossilization due to its lower temperatures.

Thus, the story of Denisova's ancestors in Eurasia began with the great climatic change of eight hundred thousand years ago, the consequences of which were far greater for their European descendants stuck in western Eurasia than they were for their Asian descendants, who were able to circulate as they pleased throughout the East, keeping nice and warm during even the worst of the cold periods. But how exactly did this common ancestor progress to these new lands in the East? Let's find out.

A *Homo heidelbergensis* learning to master fire, a skill that
Homo neanderthalensis would later perfect

10

To the East

_____

As children, both of us loved nature and spent a lot of time observing it. One thing we were particularly amazed by was the great resemblance between African and Indian animals. Elephants, monkeys, buffalos, lions . . . the territories of several species stretch across both continents, spanning the immense, arid region that stretches from the Arabian desert to the gates of India. This inhospitable zone should have acted as a barrier for these animals, yet this was not always the case. With Iran to the west and the Indian subcontinent to the east, the historical region of Balochistan is one of those arid, dusty lands where date palms and dromedaries thrive. What did African fauna do? Find a secret way of transporting food and water across this thirsty land? In reality, these lands have only been inhabitable for a few thousand years. Balochistan saw regular rainfall; its desert-like character is the result of the pressure exerted on its environment by its populations, starting in Neolithic period ten thousand years ago.

Before that, for 3 million years, the region's terrestrial climate remained rather stable and warm on average, even during glacial oscillations. When the monsoon traveled far enough, a semi-arid tropical climate developed in the south of the Arabian Peninsula, which then became studded with lakes. As a result, a Sahelian-type tropical corridor regularly opened up between Africa and India, allowing a continuity in the fauna to be established as they crossed from time to time, facilitated by the lowering of the sea level. During glacial periods, the ocean often shrunk by more than 160 feet (50 m), and the Persian Gulf (which has an average depth of around 160 feet) partially dried up. The Levant, and more generally the Near East, thus appears to have been the hub from which animal species—including generalist human forms—were introduced and then spread to Eurasia, to both the north and south.

If Elephants and Lions Could Talk

Elephants and lions are among the species that once lived in Africa, Persia, India, and Europe. Eight hundred thousand years ago, as they wandered across the vast Eurasian continent, they would sometimes have seen curious bipeds with large heads passing by: the third great human dispersal. There were two possible routes for these humans: one to the north and the other to the south.

The northern route went through Syria, Anatolia, and then, once over the Caucasus, the Pontic steppe (the corridor between the Black Sea and the Caspian Sea), leading to an immense grassy strip that linked Europe and Asia like a sort of lush highway. During very cold periods, its western branch connected the Levant directly to Europe via the frozen Bosphorus Strait, a phenomenon that has been observed many times over the course of history.

All the available evidence—depth of the strait, strength of ocean currents, fossils, differences in fauna—suggests that the northern route did not pass through Gibraltar. If the *Homo rhodesiensis/heidelbergensis* clans did take the northern route to Europe, they went via the Caucasus. The southern route reached the

Arabian Peninsula via the Nile Valley and the Levant or the Bab-el-Mandeb Strait, then southern Mesopotamia (the modern-day Persian Gulf), Eastern Iran, Southern Pakistan, and the Indian subcontinent, before continuing to the Indochinese Peninsula, Southern China, Southeast Asia, and even Australia. Although the steppe highway links western and eastern Eurasia, the temperate Far East has always been settled mainly from the south, since tropical humans tend to stay in warmer climates. What's more, the Himalayas, the Taklamakan and Gobi deserts, and the cold climate of Siberia each constituted significant obstacles, making the great plains of Northern China more accessible on foot from the south.

En Route to India

The earliest traces of the third wave of generalist humans outside Africa can be found in India. Mode 2 (Acheulean) tools were in use in the subcontinent for almost as long as in Africa, starting from 1.6 and 1.8 million years ago respectively. However, we would expect that, as in the Levant and Europe, the arrival of these humans in India around 800,000 years ago would be marked by evidence of more evolved Acheulean industries. This has indeed been the case for sites in central India: The hand axes found there are much smaller than those dating back over a million years, and comparable to European hand axes from the same period.

Perhaps the most interesting information we have, however, comes from the Sivalik Hills, a mountain range produced by the tectonic compression of the Indian plate against the Asian plate in the foothills of the Himalayas. The erosion of these mountains has revealed sedimentary strata that can be roughly dated by reversals of the Earth's magnetic field, which indicate they are between 500,000 and 1 million years old. Some of these have yielded evidence of an evolved Acheulean industry known as Soanian (after the Soan River) from that time. One of the most prolific of these sites is Atbarapur, where over 50 artifacts have been discovered. Even if the prehistory of the overpopulated continent of India remains largely unresolved, there are traces of evolved Acheulean industries in the Himalayan foothills between one million and a half million years ago, dates that coincide with the departure of *Homo rhodesiensis/heidelbergensis* from Africa. Conclusion: The third wave of generalist humans did indeed pass through the foothills of the Himalayas.

The Narmada Riddle

A fossil found in 1982 in the Narmada valley by geologist Arun Sonakia, in Madhya Pradesh, central India, has put paleoanthropologists in a panic. It has already been identified variously as an evolved *Homo erectus*, *Homo heidelbergensis*, and archaic *Homo sapiens*. Yet the "Narmada Man" only consists of half the right side of a skull and a fragment of a parietal bone. Its estimated cranial capacity, between 1,155 and 1,420 cubic centimeters, clearly excludes it from the classification of *Homo erectus* (*Homo erectus pekinensis* has a maximum of 1,030 cubic centimeters), which is further corroborated by its delicate bones, though it is assumed to be a female specimen. Though its age is unknown, it has been estimated to be between 700,000 and 70,000 years old, though the fauna found alongside it would place it at around 300,000 to 160,000 years old.

The discontinuous supraorbital ridge could belong to either Denisova or archaic Sapiens. From our perspective, the categorizations of *Homo heidelbergensis* and archaic *Homo sapiens* are equivalent. This is because in Africa, *Homo rhodesiensis/heidelbergensis* evolved into *Homo sapiens*, and we see that each time a generalist human form left Africa, its material culture arrived in India almost simultaneously. The earliest known archaic *Homo sapiens* (from Jebel Irhoud in Morocco) and the latest (from Es-Skhul and Qafzeh in the Levant) are dated at around three hundred thousand and one hundred thousand years old respectively, a range covering the probable age of Narmada humans. The specimen is therefore most likely to be archaic Indian *Homo sapiens*, perhaps mixed with other, even older local forms. Logically, *Homo heidelbergensis* mixed with the Indian *Homo erectus* that preceded it on the subcontinent (i.e., possibly "Indian Denisovans"), since the Adivasis, indigenous peoples from India, have more Denisovan DNA in them than other Indians. The work of paleoanthropologist Sheela Athreya from the University of Texas echoes the same deliberation between these two possibilities. For now, it's difficult to say with any certainty.

When did the third wave advance farther into southern and Southeast Asia? As early as eight hundred thousand years ago, it seems, since numerous hand axes have been found in Southeast Asia. The distance between the Far East and Africa meant that successive generalist human forms could only arrive in small groups. In Java, the only evidence we have is a hand axe found on a site at Ngebung. Geologist François Sémah of the French National Museum of Natural History, who excavated the site, dates this tool to around eight hundred thousand years ago. Gustav von Koenigswald described collecting "crates" of these artifacts in Sangiran, on the island of Java, but these tools were delivered to him by locals paid for that purpose . . . and could not be dated. As a result, there is doubt among most researchers today as to the age of Southeast Asian hand axes. Adam Brumm, from the University of Wollongong, Australia, has rightly pointed out that this majority position lacks coherence; this prehistorian, one of the top experts in Southeast Asian lithic industries, finds it regrettable that in Africa and Europe, the simple discovery of hand axes with an evolved surface shape is enough to signal cultural renewal, whereas this is not accepted in Southeast Asia.

Be that as it may, there's clearly a lack of well-dated sites in this part of the world. And, as we're about to discover, the situation is further complicated by a cataclysm that occurred just as members of the third wave of generalist humans were setting foot in the Far East, some eight hundred thousand years ago.

Rain of Fire

Everyone knows that sixty-six million years ago, the fall of a meteorite in the Gulf of Mexico played a major role in the disappearance of the dinosaurs. What if another meteorite had the same kind of effect on the human populations of Asia? Most of the third wave of generalist humans followed the southern shores of Eurasia. They arrived very quickly in southern and Southeast Asia, where, as several archaeological sites attest, they penetrated a region devastated and largely emptied of its inhabitants by a huge collision.

How do we know this? The answer lies in the discovery of the "Australasian tektites." Tektite fields are vast areas strewn with brown or green glass beads or micro-beads (known as tektites and micro-tektites) produced by a meteorite impact. Several fields are known worldwide, including one in Southeast Asia, which covers at least 10 percent of the Earth's surface from Tasmania to South China—an estimate obtained by taking into account only tektites weighing more than a gram. The percentage rises to 30 percent if micro-tektites are included. And yet this devastation from above is thought to have taken place eight hundred thousand years ago.

Where did the asteroid hit? The sheer abundance of the tektite field makes the apparent absence of any visible impact on the Earth's surface all the more surprising. Evidently, the collision took place somewhere near the China Seas, but the exact location is not clear. On a geological scale, this is a recent event (the Earth is 4.54 billion years old). Tectonic movements would have not had time to erase its traces. Therefore, the impact crater must still exist.

The first possible location is beneath the China Seas. Herman Burchard, a mathematician at the University of Oklahoma, posits that a meteorite struck around the Spratly Islands. Intriguingly, a bizarre underwater structure visible on Google Earth can be found in the area: a vast series of elliptical mounds measuring 142 miles (230 km) wide and 230 miles (370 km) long. Though the image is curious, Herman Burchard is not a geologist—and several experts who do work in that field have recently refuted his hypothesis. Too bad . . . it would have explained the collapse of a large part of the Sunda continental shelf. But alas, an observation made on Google Earth does not constitute reliable geological research.

The second possibility, put forward by geophysicists, is more widely accepted. Seismologist Kerry Sieh and his team from the Singapore Earth Observatory, believes he has gravimetrically identified an impact zone buried beneath a thick layer of lava on the Bolaven Plateau in Laos. Yet according to researchers, it only measures some 9 miles (15 km) in diameter, which seems very small given the gigantic area covered by the Australasian tektite field (see the map on page 145).

In short, only one thing is certain: Some eight hundred thousand years ago, a meteorite hit the region of the China Seas, an impact whose energy was great enough to send tektites over at least 10 percent of the Earth's surface, and microtektites over a much larger area. The impact must have propelled a very large quantity of molten crust into the air, which explains where the tektites come from. The Earth was then draped in a dust cloud resulting from the collision, which filtered solar radiation for a long time and caused the planet to cool. Thermal inertia would have triggered a climatic shift, leading to a change in glacial cycles.

This is one of the possible explanations for what is known as the Mid-Pleistocene Transition. This term refers to a mysterious change in glacial rhythms that occurred eight hundred thousand years ago and constitutes one of the greatest unsolved scientific enigmas of all time. The task of solving it is considered so crucial that astronomers, sedimentologists, geologists, and many other scientists have all put their minds to it—and come up with radically different answers. For the time being, the enigma remains unresolved, but it is likely that the asteroid that created the Australasian tektite field played a role.

Whatever the case, we know that the Mid-Pleistocene Transition led to the above-mentioned change to glacial periods. During almost the entire lower Pleistocene (2.58 million to 774,000 years ago), the length of each glacial cycle was around 41,000 years. This increased to 100,000 years around 800,000 years ago, resulting in greater glaciation and longer interglacial periods. As we have already seen, this phenomenon explains why Neanderthals were shaped by the cold through natural selection, as were northern Denisovans (albeit to a lesser extent).

The Devastation of the Far East

The meteorite that coincided with the Mid-Pleistocene Transition must also have had a major influence on the evolutionary fate of tropical Denisovans, since it likely devastated the whole of Southeast Asia. And the effect of a disaster of this scale on the prehistoric people would have been enormous. Firstly, given the scope of the Australasian tektite field, the mega-fires triggered by the heat released on impact must have charred virtually the entire forest of southern and Southeast Asia.

This would have been followed by an Impact Winter (i.e., an intense level of climate cooling lasting at least several years), thus delaying the return of vegetation and, consequently, wildlife to the region. A substantial proportion of the *Homo erectus* still living in Southeast Asia must have disappeared during the cataclysm. For those who survived, food was in short supply. Could it be that the enigmatic *Homo Floresiensis*, who lived far from the impact, is the descendant of an island *Homo erectus* that survived the catastrophe?

In any case, the third wave of generalist humans (which was just beginning to arrive) was only able to spread across southern and Southeast Asia after the cataclysm, in an ecosystem that had been largely devastated and emptied of its inhabitants. Sites in Southern China and Vietnam illustrate this perfectly.

A Barren Southeast Asia?

The Bose Basin sites are located in the Guangxi Zhuang Autonomous Province, bordering Vietnam to the north, far south of the Qinling Mountains. In this region of over 30 square miles (800 sq km), there are over eighty known sites containing stone tools, of which around twenty have been excavated. These efforts have led to the discovery of some fifteen thousand artifacts in a system of seven fluvial terraces that have been eroded by the Youjiang River. It was within the fourth of these terraces, in a sedimentary unit around 20 to 100 centimeters thick, that Chinese researchers uncovered cut stones, including basic and two-sided choppers, as well as large and small hand axes.

The team led by Yamei Hou from the Institute of Vertebrate Paleontology and Paleoanthropology (IVPP), who have excavated several sites, reports that out of 991 artifacts unearthed at three sites, 172 are Acheulean hand axes, 65 percent of which are shaped on one side only. The researchers point out that the "technical and behavioral skills" of those who left these hand axes "were compatible with those of the western part of the Old World"—i.e., with those of the western *Homo heidelbergensis* of GBY and Africa.

Strangely enough, these Acheulean tools were found within strata mixed with combustion products and tektites, and the radiometric date (determined using argon-argon dating) of the tektites—803,000 years—suggests that these are the result of meteorite impact. Just as strange: This Acheulean industry is represented in South China only in the Bose Basin, and within a limited stratigraphic interval—in a single layer of just one of the seven fluvial terraces. Yamei Hou and her colleagues explain this in the following terms: "One explanation can be found in the

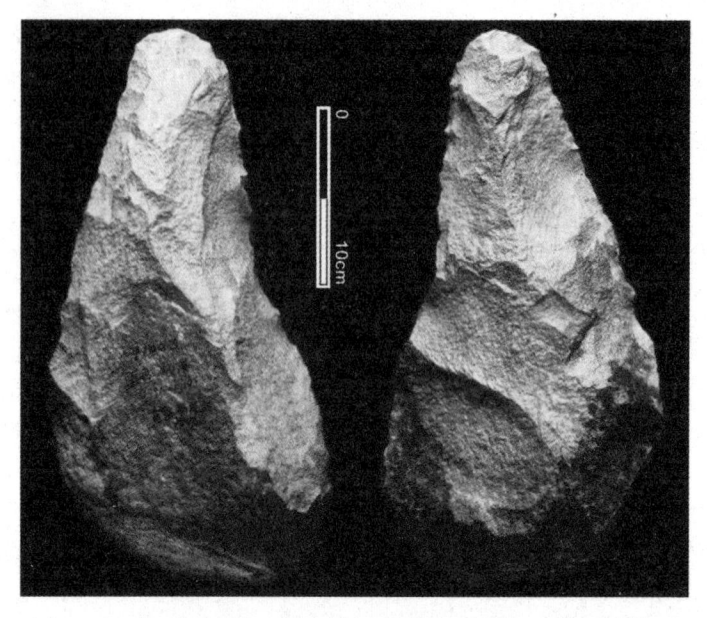

An example of a hand axe found in the Bose Basin, southern China

presence of abundant fragments of charcoal and silicified wood, detected during excavations and laboratory studies, in precisely the same sediments containing the tektites and stone artifacts. We suggest that the Paleolithic artifacts from Bose signal a behavioral adaptation to an incendiary episode, which would have resulted in the burning of all woody plants and widespread destruction of the forest, initiated by the same phenomenon that created the tektite."

In a similar case, in 2018, a Russian-Vietnamese team led by Anatoly Derevyanko of the Institute of Archaeology and Anthropology at the Russian Academy of Sciences discovered hand axes, picks, scrapers, choppers, and cutting tools at An Khê, including denticulate and serrated tools. These tools come from several sites, one of which contains a stratum with a high level of carbonization, suggesting a violent fire. This stratum also contains tektites dating back approximately eight hundred thousand years, potentially linking it to the meteorite cataclysm that occurred around this time.

The IVPP prehistorian Yamei Hou, who excavated and studied
the site in the Bose Basin

The Acheulean sites in the Bose and An Khê basins confirm that members of the third wave of generalist humans did indeed arrive in the Far East, and that, as Yamei Hou writes, "in their deforested environment" they employed lithic techniques "comparable with those of their western counterparts of the same period." These sites, which are all linked to the same time window (around eight hundred thousand years ago) just so happened to open up at the moment the third wave of generalist humans arrived.

A few examples of tektites (i.e., meteorite glass beads) found in the Bose Basin, projected by the impact from southern Asia to Tasmania

The Wave Swallowed Up by the Forest

These sites, likely dating to just after the meteoric cataclysm, are the only known and well-dated Acheulean habitats in southern and Southeast Asia, where *Homo heidelbergensis* ended up. Both are equipped with hand axe, but they are the only ones with this maker of the third wave of generalist humans. Why is it so? Anatoly Derevyanko has put forward one explanation: It is likely that humans in An Khê had to adapt to an absence of large sources of high-quality raw lithic materials, prompting them to use organic materials, such as bone, wood, and particularly, bamboo.

This is a highly pertinent observation, given that cultures across Asia have always made more tools from plant rather than lithic materials. And this changes everything in terms of available methods for tracking the third wave of generalist humans in the Far East.

In Asia, bamboo has always been a universal material used for making a variety of tools, starting with weapons.

11

The Bamboo Empire

When it comes to tracing tool industries in the Far East, paleoanthropologists have found themselves feeling well and truly . . . bamboozled. But when you think about it, we're so used to bamboo we barely even notice it: In every historical Asian film, bamboo is used to make baskets, mats, screens, furniture, and thousands of other everyday objects. It's everywhere, and has been since the dawn of time.

Bambusa vulgaris, *Phyllostachys edulis*, and *Phyllostachys bambusoides*, each from the huge *Bambusoideae* family, all grow in Asia. If modern societies can do so much with bamboo—including eat it—these plants no doubt changed the behavior of every new generalist human form that found its way to Asia. The sites of the Bose Basin and An Khê are the exception: These show evidence of the behaviors of bands of hunters who had arrived in a devastated Asia, largely devoid of its forest and bamboos, where deposits of cuttable rock could be discovered. But it did not take long for the forest to return, at which point cut stones ceased to be used for spears. They had better uses for them. . . .

This boat, made of huge bamboo rods, carries a basket also made of bamboo parts. It illustrates the wide variety of objects made from this material in Asia.

We can find out what by analyzing a site that was occupied by hunters belonging to the Sapiens species. As these humans lived more recently, they are easier to trace than *Homo heidelbergensis*. And no one can doubt the technical abilities of our species. Let's take a trip to the Laang Spean cave, occupied between 71,000 and 3,300 years ago. The Sapiens hunters who lived there used rudimentary methods: They produced very simple, single-sided tools corresponding to Mode 1 of Clarke's classification. This is surprising for *Homo sapiens*, otherwise known for their sophisticated technologies. Was this a case of regression? This cave is not the only one of its kind—far from it. Across Southeast Asia, the same crude cutting techniques can be found at numerous sites, despite the fact that members of our species across the globe were capable of shaping tools with exquisite precision by this time. Everywhere, except in Southeast Asia. . . .

But there's a simple explanation for this strange phenomenon. Hubert Forestier of the National Museum of Natural History in Paris, who excavated Laang Spean, realized that on the whole, the aim of these humans was simply to obtain sharp edges from pebbles or lithic flakes. His hypothesis is that these edges were then used to sharpen bamboo knives and tips. He points out that in Southeast Asia, the distribution area of the kinds of tools crudely shaped by *Homo sapiens* from stones or lithic flakes overlaps with areas in which bamboo are found.

The Bamboo Hypothesis

Hubert Forestier therefore came to the same conclusion as Anatoly Derevyanko when the latter analyzed the An Khê sites. Bamboo, Hubert Forestier points out, "thanks to its exceptional 'solid-flexible' qualities, can be used as a material for making formidable tools, utensils, and weapons." It therefore makes sense that this material replaced stone in prehistoric Asia. This theory, known as the "bamboo hypothesis," explains why the lithic industries of the southern Far East remained so underdeveloped. If this was true at the time of the arrival of *Homo sapiens*, it certainly would have applied eight hundred thousand years ago, at the time of the arrival of *Homo heidelbergensis*. And, like our own species, members of the third wave of generalist humans found an infinite number of technical solutions in bamboo.

The bamboo hypothesis is all the more convincing in that it can be applied to all waves of generalist humans, even if evidence of the transition from stone to bamboo tools is only clear in the case of the last two: *Homo heidelbergensis* and *Homo sapiens*.

In the case of the latter, anthropologist Wilhelm Solheim (1924–2014) was well aware of the phenomenon, for which he coined an interesting neologism: "lignic." This term, inspired by the main component of wood, lignin, is effective because it resembles the term "lithic." It therefore suggests that bamboo played the same universal role in eastern Eurasia as stone played in western Eurasia.

This welcome neologism is an efficient term for characterizing all prehistoric Asian cultures, in which techniques using plant-based materials—including bamboo—far outweighed those using stone. The latter still played an essential role, since it was used to cut plants, including bamboo. It would be excessive, therefore, to describe cultures as exclusively lithic or lignic, seeing as all prehistoric cultures were a combination of both.

For example, in the case of *Homo heidelbergensis*, the ancestor of Denisova, the remains of a plank of wood at GBY in Israel and a wooden structure dating back 476,000 years ago at the Kalambo Falls site in Zambia are evidence of wood construction having been practiced by members of the third wave of generalist humans.

In western Eurasia, the aforementioned wooden spears from Schöningen, Germany (two hundred thousand years old); the spear from Clacton-on-Sea, England (four hundred thousand

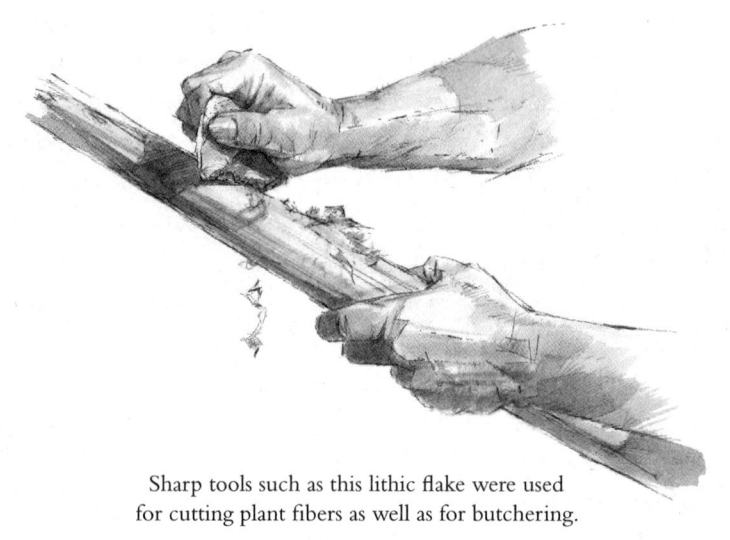

Sharp tools such as this lithic flake were used
for cutting plant fibers as well as for butchering.

years old); and the digging sticks and other Neanderthal tools discovered at Barrika, in the Spanish Basque Country (ninety thousand years old), all illustrate the use of wood by *Homo heidelbergensis* or their Neanderthal descendants.

The Asian Swiss Army Knife

In the great forest of tropical Asia and in temperate Asia, by contrast, the role of wood was so prominent that after the arrival of each wave of generalist humans, stone-cutting regressed drastically to make way for lignic industries. The *gramineae* family (woody grasses with a tree-like growth form) to which bamboo belongs, had many uses. *Bambusoideae* (the bamboo subfamily) is very diverse, and once made up a rich, immense forest stretching from the Wallace Line up to the latitude of Korea toward the sea and reaching up to 12,500 feet (3,800 m) in altitude. There are two main families of bamboo species that are known: One is adapted to tropical climates and the other to temperate climates. To this day, China holds the title of "bamboo country" as the country with the largest bamboo forest in the world. The plants are so ubiquitous in Asia that the native bears—pandas—have adapted to eat their leaves. While the same does not apply to humans, they do eat their shoots and have taken advantage of bamboo in all its shapes and sizes, utilizing stems as thin as fingers or wide as tree trunks for various purposes. As these plants grow fast, they provide an abundance of material and are always on hand, unlike high-quality rock deposits, which are scarcer in Asia. For example, flint deposits, the stone of choice for Europeans, are only commonly found in northwestern China, far beyond the Qinling Mountains.

As a result, Asian prehistoric peoples fashioned almost all their tools from bamboo. Tools made by modern hunter-gatherers in Highland Papua give us a good indication of what these might have looked like: knives, kitchen utensils, musical instruments, ties, baskets made with strips, racks, traps, and even entire houses. Every family has their own traditions and expertise in different varieties of bamboo. As for weapons, Papuan hunters fashion spears, bows, arrows, and blowpipes, as well as a host of traps using cages, spikes, and other mechanisms. All of this expertise is kept secret within each family and passed on from father to son, or even mother to daughter.

A Sino-American team led by Harvard University prehistorian Ofer Bar-Yosef (1937–2020) has put the bamboo hypothesis to the test. The team fashioned several bamboo tools and successfully demonstrated that it would have been possible in a prehistoric context to use this material in a number of ways. However, the researchers' experiments also highlighted the limitations of bamboo blades—e.g., for cutting large animal skins, which a stone blade can do perfectly. In short, the exact balance between lithic and lignic tools in Asia or Europe is not yet fully understood.

All these observations do however lead us to a simple conclusion: While there's no doubt a third wave of generalist humans came into China

Following a brilliant career at the Hebrew University of Jerusalem, the prehistorian Ofer Bar-Yosef joined Harvard University.

from the south, the absence of Acheulean cultural markers means we don't know a lot about them. Limited evidence of lithic industries is our only way of tracing their footsteps through Asia, and as we know, our knowledge of their existence before a shift toward lignic industries (at Bose and An Khê) was a lucky accident made possible by the giant meteorite. With this in mind, we can look at the progression of these humans throughout continental Asia. Let's start by taking a look at some of the characteristics of this region.

China: The Middle Kingdom

Zhōngguó, the "middle kingdom" is one of the names for China in Mandarin. Looking at the geography of continental Asia, it's easy to understand why inhabitants of this part of the Earth might have felt as though they were at the center. Their surroundings consist of two vast coastal compartments, one to the north and the other to the south, separated by the Qinling Mountains (also known as the "Ash mountains"). To the west lies the immense Tibetan plateau, a largely uninhabitable area given its average altitude of almost 15,000 feet (4,500 m). It is flanked to the north by the Tarim Basin and the Gobi Desert, vast strips of extremely dry desert land where huge fluctuations in temperature make for hostile living conditions. The region is also home to the Loess plateau, the largest in the world of its kind, which flanks the Gobi Desert and the Tibetan Plateau to the west of the eastern plain. While desert-like, it is potentially fertile. Huáng hé—the Yellow River—emerges from here, colored yellow by the Loess Plateau, and irrigates the northern continental plain, the cradle of Chinese civilization.

To the southeast of the Loess Plateau, the Qinling Mountain range stretches over 930 miles (1,500 km) from east to west, outlining the northern limit of the monsoon region. China's most important river—the Yangtze River or Blue River—also runs through these mountains.

South of this barrier, the climate is humid and tropical, tending toward oceanic near the sea, and the geology is dominated in many parts by the world's largest and most diverse limestone landscapes. A huge drainage basin of almost 125,000 square miles (200,000 sq km) is crisscrossed by short, fast-flowing rivers, whose water comes from monsoons and typhoons.

Some 40 percent of this vast area is dominated by mountains and hills over 6,500 feet (2,000 m) high. This biogeographical structure has always had a major influence on the evolution of populations by consistently bringing together populations from a large tropical basin with those from the northern basin, which could then be repopulated after cold periods.

In China, the Evidence Is Far from Cast in Stone ...

Did these humans move northward like all tropical humans before them? Unfortunately, their occupation of the southern compartment is difficult to analyze from the archaeological data available, since fossilization of both human remains and lignic artifacts is rare in tropical environments. Still, it's logical to assume that after their arrival around eight hundred thousand years ago, *Homo heidelbergensis* spread very rapidly, as far as the Qinling mountains. Members of the third wave of generalist humans would likely have found themselves blocked by this geographical barrier marking the limits of tropical Asia, which would have stalled their progress before they continued to venture northward.

As we have already seen, the Acheulean lithic industry (hand axes) cannot help us trace their footsteps, since it is absent from mainland China, only reappearing at the western end of the Qinling Mountains 250,000 years ago, well after the arrival of the third wave. What can be found throughout continental Asia, are the so-called lithic core flakes we saw in the bamboo hypothesis, the crude shaping method likely used by all human waves. Even if flint deposits were accessible in the north, there's no doubt that *Homo heidelbergensis* adopted the efficient plant-based industries already practiced by local populations they encountered and mixed with. All these reflections lead us to a simple conclusion: While there is hardly any doubt that the third wave of generalist humans arrived in continental Asia from the south, there are no biological markers (fossils) or cultural markers to provide evidence of this, except for one—fire.

Fire in China, Fire Everywhere

As in Europe, evidence of fire in Asia between 800,000 and 400,000 years ago is either absent or highly uncertain. However, layer 10 at the Zhoukoudian (Peking Man) site and the notes of its first excavators suggest that fire was already being used in the Nihewan Basin some 770,000 years ago. This hypothesis is compatible with the GBY study team's impression that fire had been used intentionally at the site, but theory is not widely accepted by paleoanthropologists. On the other hand, traces of fire dating back less than four hundred thousand years are completely accepted as evidence of these behaviors. This date marks the beginning of widespread domestication of fire throughout the Old World, and therefore also throughout Eurasia. This evolution, as we have seen, corresponds with the arrival of third wave generalist humans with more developed cognitive abilities.

To conclude, as in western Eurasia, it's plausible that fire was used in a controlled way in eastern Eurasia as early as eight hundred thousand years ago (Zhoukoudian layers 7 and 8). As in Europe, we can be certain it was used as early as four hundred thousand years ago (Zhoukoudian upper layers, Fenhe Basin)— i.e., at the time of the emergence of Denisova.

The Chronology of Denisova

Our journey through the history of *Homo heidelbergensis* ends here, but before going any further, we need to summarize it. The third wave of generalist humans emerged from Africa around eight hundred thousand years ago, then conquered territories from Europe to the Far East. They first followed the southern shores of Eurasia, soon reaching India and then Southeast Asia. They then stopped south of the Qinling Mountains, adopting a way of life that revolved around the use of bamboo.

They eventually conquered northern Asia, though this occurred later due to the constraint of the cold. As glaciers advanced into northern Asia and the Tibetan plateau, the possibility remained of joining populations living in the milder climates of the south, or even in the tropical belt. This means the demographic effects of the ice ages in the Far East were far less severe than in Europe, where only small Neanderthal populations highly adapted to the cold were able to survive. In northern Asia, after severe climatic episodes, populations from the large tropical basin could come to reinforce those that had remained in the north. Descendants of the third wave of generalist humans would become Denisovans in eastern Eurasia, and Neanderthals in western Eurasia. These sister groups met in central Asia on the steppe highway (notably around the Denisova Cave) and interbred there.

Yet these evolutionary differences due to geography do not preclude a certain chronological parallel existing between western and eastern Eurasia. Having arrived in the southern regions of the Far East and Far West respectively around eight hundred thousand years ago, members of the third wave of generalist humans interbred around the same periods of time with the "super-archaic" humans they found there. Each then went on to their respective part of the world, where they evolved in parallel to one another. This implies that the chronological reference points we have for the evolution of Neanderthal—the best-known Eurasian species after our own—also apply to Denisova, the least-known Eurasian species.

It was therefore from 400,000 years ago onward, when Neanderthalization was in full swing in western Eurasia, that the process of "Denisovation" was happening in parallel in eastern Eurasia. Like the Neanderthal era, the Denisovan era began around 300,000 years ago and ended around 40,000 years ago after the arrival of *Homo sapiens*. This timeline is confirmed by the dates of inhabitation of the Denisova and Baishiya caves (the only caves widely accepted to have been home to Denisovans) at the eastern end of the Tibetan plateau. In the sediments of the Denisova Cave, the oldest traces of Denisovan mitochondrial DNA are over 200,000 years old, while the most recent are around 40,000 years old. At the Baishiya site, on the Tibetan plateau, traces of the same DNA in the cave sediments range from 160,000 to 45,000 years old.

This clear new paradigm, suggesting that Denisova was to the east what Neanderthal was to the west, provides a helpful method for tidying up Eurasia's Pleistocene fossil record. A lot of confusion is caused by the resemblance between the Eurasian fossils: All bear archaic traits derived from their common ancestor, *Homo heidelbergensis*, which makes them similar in appearance. Their differences, meanwhile, can be explained by the fact that they evolved in separate environments and, above all, in different demographic situations: The Neanderthals remained in isolation, but the Denisovans did not. With this information in mind, we can now draw up a portrait of Denisova.

Map of China indicating the location of pre-Denisovan and Denisovan fossils

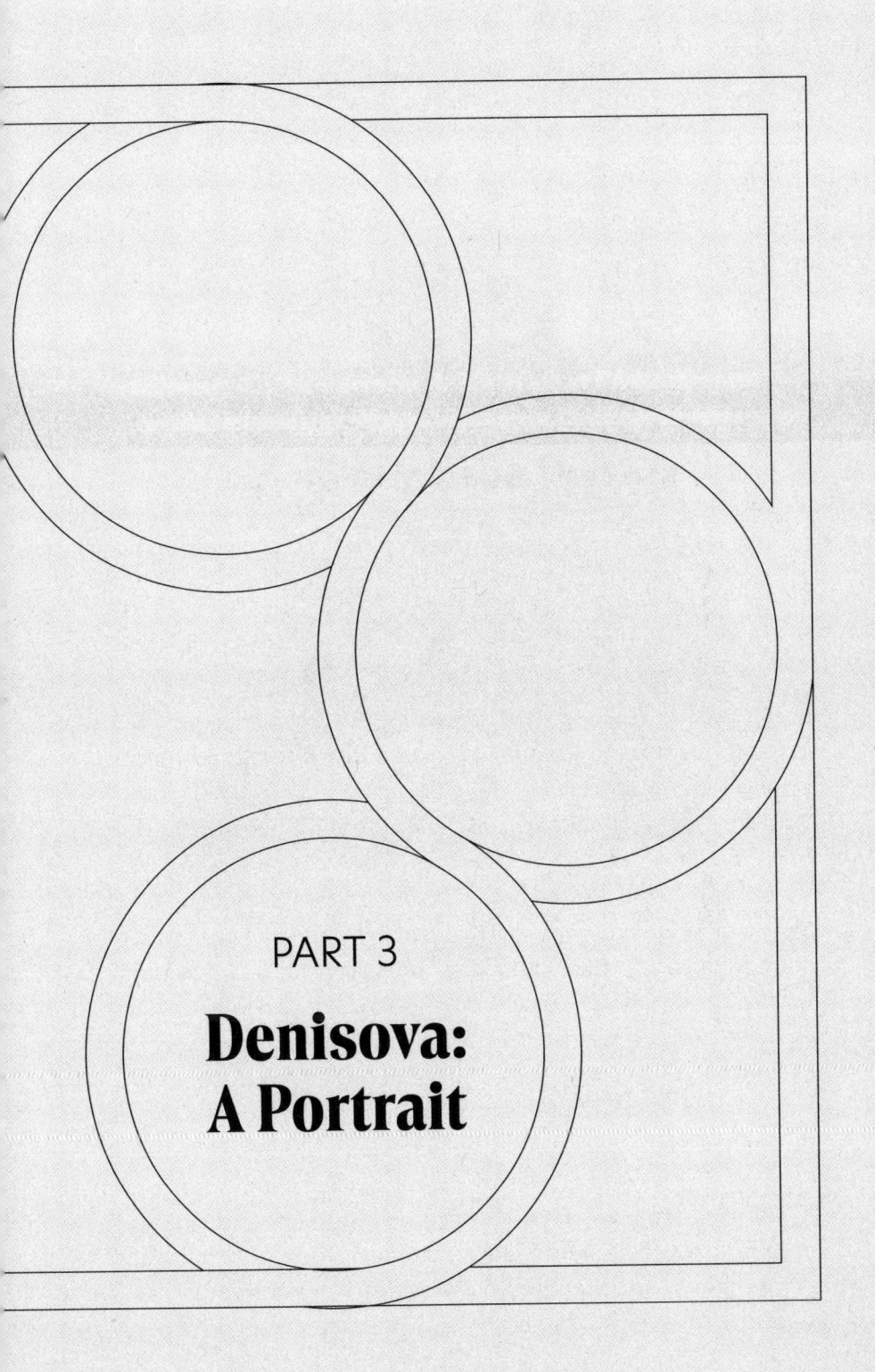

PART 3

Denisova: A Portrait

This Denisovan resembled his Neanderthal siblings,
only with Asian characteristics.

12

An Identikit Portrait

Our investigations into human evolution in Asia were preceded by the explorations of a certain Belgian adventurer whose findings were of particular interest. As told in the famous story *Tintin in Tibet*, the intrepid reporter left for Tibet on a hunch, to help his friend Chang. In the mountains of this immense region, he discovered a very interesting form of hominin: a Yeti. But the "Abominable Snowman" did not quite live up to his name. In fact, he was a sensitive, solitary being, who saved Chang and was sad to say goodbye to the friend who came to his rescue: Tintin. From a scientific point of view, the Yeti is a cryptohominoid (i.e., a mythological innovation), known from Asia to North America. It acts as a psychological tool to help us untangle the question of our origins by imagining an ancestral form that is half man, half monkey: hairy, primitive, enormous, strong, and endowed with human feelings. The discovery on the Tibetan plateau, not of the Yeti, but the "abominable snow Denisovan" was therefore an astonishing one indeed.

But in order to find Denisova just as Tintin found Chang, we need to find fossils—and we need to make them talk. The tiny fossils discovered in the Denisova Cave are not enough, so we now have to go in search of fossils where the Denisovans lived: all over Asia. As we'll see over the next chapters, China's excellent prehistorians have inadvertently accumulated quite a collection of fossils that are no doubt Denisovan. But before we begin to scour the lands of Asia, we need a forensic reconstruction of the subject. To paint this portrait, we have two crucial pieces of information at our disposal. The first is the only Denisovan fossil that has consistently been acknowledged by paleoanthropologists as such: the Xiahe mandible. The other comes from new genetic studies that can tell us about the appearance of our Asian sibling. Like scientific detectives, we are going to create a surprisingly detailed forensic reconstruction of Denisova.

Buddha and Denisova in the Mountains

The main feature of our portrait is a discovery worthy of its own adventure film. It's a bone that was found in the Baishiya cave sanctuary in central China, where Tibetan Buddhists go to meditate. You might be wondering how Tibetan Buddhists found themselves in central China. Even if the Western world is generally unaware of it, Tibetan civilization extends across the whole of the "Tibetan plateau," a vast steppe at an average altitude of almost 15,000 feet (4,500 m), stretching from the Himalayas to the heart of China. The plateau is populated mainly by ethnic Tibetan Chinese, but also by Turco-Mongol and Han ethnic groups. The Baishiya cave is located near the town of Xiahe, in Gansu province, in the far northeast of the Tibetan plateau. Labrang Monastery, the most important center of Tibetan Buddhism outside the Tibetan Autonomous Region, is located at Xiahe, oddly enough, in central China.

This combination of geographical and cultural particularities explains why this cave, perched at an altitude of 10,764 feet (3,281 m), is also a holy place where followers of the Tibetan rite have long come to retreat and pray. It was in the cold depths of the cave that, in the 1980s, a monk discovered half of a fossilized mandible. He handed it over to the Sixth Gung Thang Lama, who could easily have considered grinding it into powder for medicinal purposes. Fortunately, the religious dignitary kept the specimen and handed it over to a geologist at Lanzhou University, the great university of Gansu.

In July 2016, Jean-Jacques Hublin, then at MPI-EVA, received a series of photographs from his colleague Zhang Dongju, of Lanzhou University. These included what paleoanthropologists now call the "Xiahe mandible." In an article for *Pour la Science* detailing his discovery, he noted: "This e-mail was the starting point for a most exciting collaboration with Zhang Dongju and his colleague Fahu Chen. The specimen was clearly not Sapiens. It was very robust, did not have a strong chin and bore huge teeth."

As luck would have it, Xiahe's mandible was encased in limestone ore rock, which could be dated using the uranium–thorium radiometric method to some 160,000 years ago. This placed it at the height of the Denisovan Age, during a mild period. The question was whether or not the fossil itself was Denisovan.

After IVPP's unsuccessful attempt at DNA extraction, Zhang Dongju and Jean-Jacques Hublin's team turned to another identification technique based on proteins (the body's workhouse molecules, whose blueprint is inscribed in DNA). The proteome—i.e., all the body's proteins—can be used in the same way as the genome to identify a species. And it has a clear advantage over DNA: Certain proteins, such as those making up collagen—one of the components of skin and bone—have a much longer lifespan than genes. In 2019, after extracting collagen from the mandible, Frido Welker from the University of Copenhagen was able to establish its protein structure. It turned out to be very similar to the protein structure encoded in the DNA of the *Denisova 3* fossil.

For the first time since 2010, a new Denisovan fossil had finally been identified. What's more, it had been found in central China, over 1,200 miles (2,000 km) from the Denisova Cave, where all fossils are almost automatically deemed to belong to the *Homo erectus* species. Despite this evidence that the Xiahe mandible was Denisovan, doubts remained. As mentioned in chapter 4, a team led by Zhang Dongju set out to find Denisovan DNA in the sediments of the Baishiya cave, which they did. This time, it was certain: A mandible had been found that belonged to Neanderthal's Asian sibling. Today, along with the jaw of Penghu, this is the only accepted Denisovan fossil aside from the micro-fragments and macro-teeth discovered in the Denisova Cave.

The Abominable Snow Denisovan

What does this fossil tell us? Its strong and robust morphology is similar to that expected of Middle Pleistocene human species: *Homo erectus*, *Homo heidelbergensis*, and their descendants Neanderthal, archaic Sapiens, and Denisova. The robust structure, length of the body, and lack of chin that characterize this Denisovan mandible are features associated with archaic humans. In particular, the mandibular foramen—a hole in the mandibular wall that allows nerves

Members of Zhang Dongju's team collect sediments in the Baishiya cave for DNA extraction.

and blood vessels pass through—is found beneath the second premolar, as in Denisovans' ancestors *Homo heidelbergensis*.

Neanderthal should also have had this archaic feature, especially as it is present in Sapiens. But its isolation, exacerbated by the ice ages, had subjected the species to strong genetic drift in Europe: In this case, changes to the shape of the face shifted the mandibular foramen beneath the first molar. This position of the Denisovan foramen tells us that, unlike the "snout-like" Neanderthal face, the Denisovan face is less forward-projected and therefore flatter. This brings it closer in appearance to our own *Homo sapiens* face, with the area of the chin and beginning of the forehead in the same vertical line. This detail is crucial, as it is one of the reasons why some Chinese paleoanthropologists have long believed that *Homo sapiens* may have originated in Asia.

However, like Neanderthals, Denisovans did not develop a "bony" chin. Instead they retained the inwardly receding chin of *Homo heidelbergensis*. The precise morphometric comparison carried out by the research team confirms that the general characteristics of Xiahe's mandible are similar to those of other known mandibles from the Middle Pleistocene, even if the shape of its dental arch is closer to that of Neanderthal than *Homo erectus*, and further from that of Sapiens.

Some 160,000 years after the death of the Xiahe Denisovan, its mandible still has two molars intact. The morphology of these is also in line with that of Asian human teeth since the arrival of the third wave of generalist humans in Eurasia. They have five principal cusps (i.e., five well-developed dental eminences), plus two accessory cusps. As a result, the occlusal surface the grinding surface of the tooth—is very large. This feature of Denisovan teeth, already noted on the molar discovered in the Denisova Cave, is consistent with a diet largely consisting of plants, a tendency already noted on certain *Homo erectus*.

X-rays have also revealed very wide roots. The second molar has three roots, as does the only known Denisova tooth. This feature is crucial, as it is present in only 3.5 percent of non-Asian

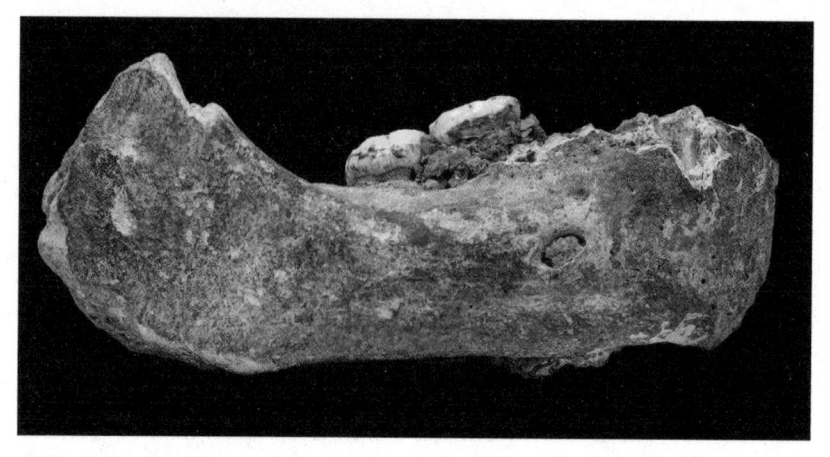

The Xiahe mandible was for many years the only accepted Denisovan fossil, apart from the tiny phalanx and enormous molar found in the Denisova Cave.

Sapiens, but in 40 percent of Chinese and Native Americans. Paleoanthropologist Sarah Bailey, from New York University, considers the three roots of the second molar to be a Denisovan trait in their own right, acquired by Asian Sapiens through interbreeding with Denisova.

What kind of sketch does the Xiahe mandible paint for us? That of a human with a powerful jaw, endowed with substantial teeth, a receding chin, and a face that is less prognathic (i.e., does not protrude in relation to the forehead) than that of a Neanderthal.

But there are other traits we can deduce from the specimen. Since Denisovans are descended from *Homo heidelbergensis*, we would expect them to have a cranial capacity at least as large as that of their ancestors. Denisovans' brains continued to evolve, meaning their endocranial volumes were likely comparable to those of Neanderthal and Sapiens, somewhere between 1,100 and 1,800 cubic centimeters (with a wide margin), a range with a median value of 1,350 cubic centimeters.

In short, Denisova was a human with a powerful jaw, large teeth, and presumably a large head to go with it. Beyond this we can only speculate. Since the mandible is primitive in appearance, the face likely was too. It's reasonable to imagine they also had massive superciliary arches, a trait shared by all human forms except *Homo sapiens*. Since the mandibular muscular attachments are strong, they are reminiscent of those of Neanderthal, who also possessed strong temporal attachments, along with a low, backwardly elongated skull as in most ancient humans. Still, with the information we have so far, we wouldn't have much chance of picking Denisova out of a lineup if this were a detective film....

Hexian, Chaoxian, Dongzhi... the First Denisovans?

Large teeth that have been discovered, comparable to those of the Xiahe mandible only much older, suggest that Denisovation was already underway around 400,000 years ago. The first of these teeth are part of the Hexian fossils from the Longtan Cave on Wanjiashan Mountain, which overlooks the Blue River just north of the Qinling Mountains. The "Hexian fossils"—i.e., "fossils from the He district"—were discovered by chance in 1981 and include two cranial vertexes, cranial fragments, a fragment of a left mandibular body with the second and third molars attached, and ten loose teeth. After much discussion, these fossils were dated at around 412,000 years old.

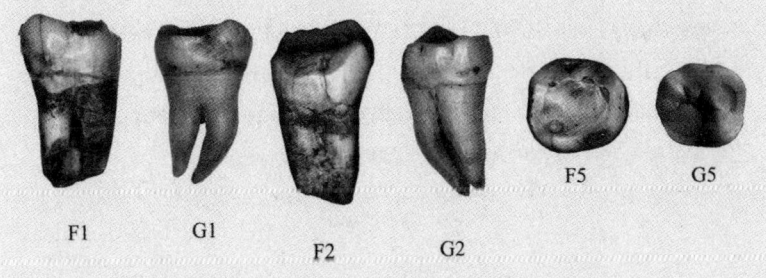

The fossilized teeth discovered in Dongzhi indicate macrodontia among the individuals to whom they once belonged.

This date, which is widely acknowledged today, is close to the approximate beginning of Denisovation, suggesting the Hexian human was pre-Denisovan. Its endocranial volume of almost 1,025 cubic centimeters verges on the lower limit for a species to be classified as Denisovan: 1,100 cubic centimeters. Its large teeth appear consistent with the Denisovan characteristic of macrodontia. The findings on the Hexian fossils therefore support the idea that Denisovation began at the same time as Neanderthalization.

This is further confirmed by the three large teeth attached to the "Chaoxian maxilla," a fossil presumed to be 310,000 years old, discovered in the 1980s near the village of Yin Shan, in Chaoxian, Anhui province. These are 12 to 13 millimeters wide, while modern human teeth are 6 to 8 millimeters wide. Could these be *Homo erectus* teeth? No. Given their age, these are pre-Denisovan teeth.

The most substantial pre-Denisovan fossils were discovered in Dongzhi. The fossils, found in the Hualong Cave, Anhui province between 2000 and 2010, comprise 30 or so human skull fragments that undoubtedly belong to a single individual, with the possible exception of a lower molar. The individual has been painstakingly dated to "over 300,000 years ago" on account of its estimated cranial capacity of around 1,150 cubic centimeters, meaning the Dongzhi human was most likely pre-Denisovan. This estimation is further reinforced by the size of the lower right second molar, which has seven cusps (reliefs), just like the Denisova and Xiahe teeth.

In short, the Hexian, Chaoxian, and Dongzhi fossils all show signs of macrodontia like the Denisovans who came after them, suggesting these ancient humans were pre-Denisovans.

Someone Call the Science Police

Where do we go from here? It's time to call the forensic detective team, or more specifically, the geneticists who are able to complete the picture for us. In an unprecedented feat, they were able to deduce from the Denisovan genotype many of the apparent traits of the Denisovan phenotype. Thanks to the knowledge that has been accumulated on genome expression, it is now possible to reconstruct the broad outlines of an individual's physical traits, including their face, head and body shape. The color of our eyes, hair, and skin, but also the shape of our nose, the size of our ears, and the width of our forehead, are all recorded in our coding DNA—i.e., in the 20,000 or so genes that represent 1.1 percent of our DNA.

Liran Carmel and his team from the Hebrew University of Jerusalem carried out this exercise using the perfectly sequenced genome of *Denisova 3*, some of whose genes have already indicated brown eyes, black hair, and dark skin. Commenting on the publication of the results in the journal *Cell*, Liran Carmel said: "This work is a step toward being able to deduce an individual's anatomy from their DNA. It is the dream of police forces the world over, but currently remains only a partial possibility."

In Denisova's case, this extremely meticulous work took the team more than three years to complete. How did they do it? The answer is a little technical: They looked at a factor known as DNA methylation. Over the course of an individual's life, each of the bases making up its DNA (C, A, T, and G) can be modified by the addition of a methyl group: –CH3. This is what is known as an "epigenetic" ("around genetics") phenomenon, as the addition of methyl groups results in blockages in the translation of genes into proteins, known as "repression of gene expression." In each species, this phenomenon affects genome expression in a specific way: Geneticists say that a certain "methylation profile" modulates gene expression.

The researchers focused on the methylation profile affecting *Denisova 3*'s observable physical traits. To do this, they compared it with the methylation profiles of Sapiens, Neanderthal, and the chimpanzee, in order to understand the specific ways in which methylation affects genome expression in Denisova. David Gokhman, part of Liran Carmel's team (who invented the

Liran Carmel, from the Hebrew University of Jerusalem, presents a rendering of the teenager *Denisova 3* made by deducing her phenotype from her genotype.

method) explains: "In this way, we can predict which parts of the skeleton are affected by the differential regulation of each gene and how this would cause that part of the skeleton to change—for example, a longer or shorter femur."

And so, the phenotype (i.e., the observable characteristics) of *Denisova 3* was put forward.

A Funny Type of Phenotype

Of the fifty-six Denisovan features identified either in relation to Neanderthal or to present-day humans, thirty-two helped the researchers to identify likely Denisovan anatomic traits (see the diagram on the next page).

To begin with, they identified archaic traits common to Denisova, Neanderthal, and archaic Sapiens. Each of these are associated with a general robustness: thick walls on the bones of the limbs, a broad facial structure that protrudes from the skull (a trait that has disappeared in present-day *Homo sapiens*), and large shoulder blades, and very dense bones. These traits are all to be expected, as they were already present in *Homo heidelbergensis*.

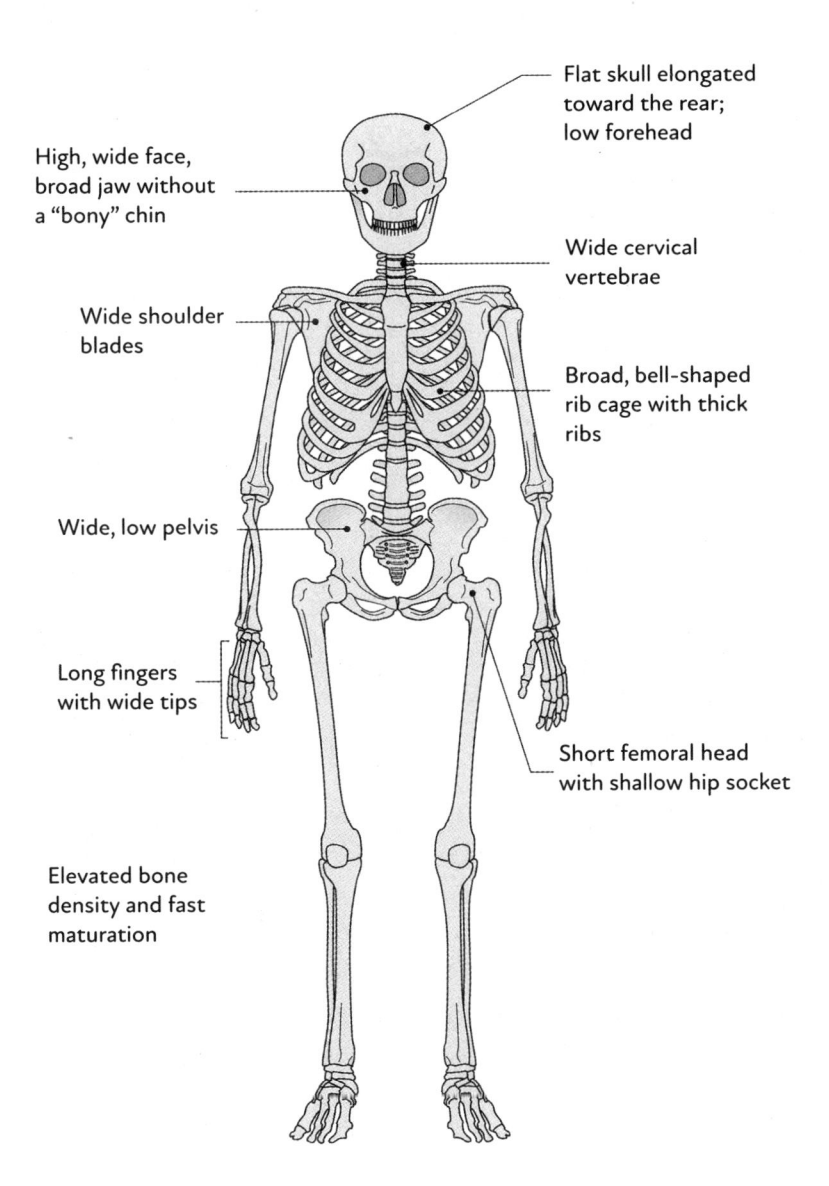

Flat skull elongated toward the rear; low forehead

High, wide face, broad jaw without a "bony" chin

Wide cervical vertebrae

Wide shoulder blades

Broad, bell-shaped rib cage with thick ribs

Wide, low pelvis

Long fingers with wide tips

Short femoral head with shallow hip socket

Elevated bone density and fast maturation

Denisovan skeleton derived from *Denisova 3*'s methylation profile

Next, the researchers identified twenty-one archaic features shared only with Neanderthal, including a robust mandible, low forehead, elongated skull, broad skull base, wide pelvis with a large femoral head followed by a short, broad femoral neck, massive phalanges, and a wide, bell-shaped rib cage. These derived traits are indicative of a robust, sturdy body built for the endurance required for walking, similar to that of Neanderthals.

There are eleven traits that differ between Neanderthal and Denisova. Some of these are specific to Denisovans, such as an elongated, U-shaped dental arch, a widened skull at the temples and top of the head, and very wide mandibular joints (condyles). Other traits are common among both Denisovans and Sapiens, such as the consistent width of the skull from the parietals to the base.

From these genes, Liran Carmel's team has produced a rough portrait compatible with the—hypothetical—one we sketched above, based on Denisova's origins, one mandible, and a few teeth. On the whole, our portrait is one of an archaic-looking human, with thick bones, wide hips, strong limbs, a rugby-ball-shaped skull, and a long face. This is consistent with the archaic features observed in the Xiahe mandible, reflected in a number of traits shared by both Denisovans and Neanderthals—traits that already existed in their common ancestor. Liran Carmel's team confirmed these same archaic traits, which they deduced from a single Denisovan genome. They also predicted derived traits for Denisova, which may or may not be shared with Neanderthal. These are linked to the ecosystems in which these two forms evolved: cold to temperate for Neanderthals and northern Denisovans, but tropical for southern Denisovans. Denisova would have developed regional adaptations, which gave it greater phenotypic diversity than Neanderthal. Clearly, Denisova and Neanderthal are Eurasian siblings, closer to each other than they are to their African cousin *Homo sapiens*.

How should we proceed with our investigation from here? We have already obtained solid chronological and geographical reference points. We've solved a substantial chunk of the puzzle of Denisova's phenotype. Now it's time to go on a fossil hunt.

Denisova's Genes and Physiology

It's now time to add physiological traits to the anatomical sketch we've drawn so far. These traits can be deduced from Denisova's genetic heritage—i.e., the genomes of contemporary Asian populations. Around 10 percent of the Denisovan genome can be reconstructed by deducing traits from Sapiens genes. The Denisovan DNA that can be found in the Sapiens genome is the result of positive selection that occurred through the reproduction of *Homo sapiens* descendants who interbred with Denisovans. We have already touched upon the EPAS1 gene, involved in the physiological response to low oxygen levels at high altitude. Other common Denisovan genes—such as FADS1, SUMF1, PLPP1, LIPA, CH25H, TBX15, and WARS2—are involved in metabolism, particularly fat metabolism. TBX15 and WARS2 present a different variant to that found in Neanderthal, which is also found in Greenland Inuit and Native Americans. These genes influence the distribution of fat in the body, particularly subcutaneous fat. These may have been involved in the distribution of fat in the face, and in particular in the lips, which suggests Denisovans may have had fleshy lips.

In addition, part of the Denisovan genome regulates the immune response to rainforest pathogens. This was retained when the first *Homo sapiens* reproduced in Southeast Asia, because it favored the species's adaptation to these new territories. Similarly, the Denisovan version of the TNFAIP3 gene that influences immune response does not suggest a strong inflammatory response, unlike the Neanderthal variant of the same gene, which many *Homo sapiens* have inherited. Incidentally, this Neanderthal gene variant was, along with comorbidity and obesity, a major genetic risk factor for COVID-19 in the Eurasians who inherited it.

This northern Denisovan (the Jinniushan
woman) was adapted to cold living conditions.

13

In Search of Lost Fossils

In 1978, the *salvador* of the Denisovan species came in the form of the Dali fossils. But the name does not refer to the Spanish painter. It refers to Dali County in Shanxi, a few around 20 miles (30 km) from the Yellow River. It was here, in 1978, that a farmer discovered a skull near the village of Jiefang. Alongside Jinniushan and Harbin, this fossil is without doubt the most important in the Chinese fossil record, and appears to be the foundation of the Denisovan lineage.

Some of the interpretations that have been made of the Dali fossil are surrealist indeed—but rather than getting too far ahead of ourselves, let's start from the beginning. Unfortunately, the farmer who discovered the Dali skull was not thinking of stratigraphic considerations, so once again, it was difficult to date this Chinese fossil (see page 198). In 1994, the IVPP made a first attempt by applying the uranium-thorium method to a bovid tooth found on the site, which was dated to 209,000 years ago. Still, as there was no guarantee that the tooth came from the same stratigraphic level as the skull, this caused an outcry. In 2017, the use of several simultaneous methods dated the fossil to around 260,000 years ago, placing Dali right in the middle of Denisovation.

The Asian/European parallel suggests that the emergence of the Denisovan lineage occurred between 400,000 and 200,000 years ago. During this period, Neanderthal traits became more prevalent in Europe, so the same is likely to apply to Denisovan traits in Asia. We will see how these came together in the Dali fossil. The fossil exhibits both archaic and derived traits that characterize it as pre-Denisovan. In terms of archaic features, this skull features thick bones marked by marked reliefs, a low, long cranial vault, and a receding forehead starting from a very large supraorbital torus (brow ridge). In terms of derived features, it has a cranial capacity of 1,120 cubic centimeters, placing it at the lower end of the Neanderthal spectrum. Even more original, its highly pronounced supraorbital bulge shrinks in the center, distinguishing it from the "visor"—the continuous torus—of the Neanderthals. Unless the deformation of the skull in sediment has misled us, the face is less forward of the cranium than in Neanderthal, but not as much as in *Homo sapiens*, as we've already seen from our initial Denisova sketch.

Otherwise, the face has a slender, flat appearance. Instead of being triangular as in *Homo erectus*, the skull is wide at the level of the parietals—the two bones at the top of the skull—and then remains the same width all the way to the base, a Denisovan feature that corresponds to the broad head shown in our sketch. As the maxilla was fractured during fossilization, the face doesn't appear to be high, but interestingly, computer reconstruction of the skull reveals that it was in fact very flat. This is a "Mongolian" feature brings to mind the faces of many Asians when compared to those of Africans and Europeans. Unfortunately, the fossil does not contain any teeth, which could have helped establish whether or not it had the characteristic Denisovan macrodontia.

Dali: Surrealist or Realist?

In 1979, in a preliminary description, Chinese prehistorians initially regarded Dali as a transitional form between the Peking

Man and Neanderthal. They gave the specimen the rather meaningless label of "late *Homo erectus*." Perhaps even more surprising is the interpretation of the Dali human by Xinzhi Wu (1928–2021), then vice director of the IVPP. Based on a full description of the specimen drawn up in 1981, this distinguished paleoanthropologist found that Dali's face possessed a transient morphology between Peking Man (*Homo erectus pekinensis*) and *Homo sapiens*. For him, this heralded the emergence in Asia of a "Mongoloid" form of *Homo sapiens*. Xinzhi Wu did not deny that Sapiens migrated out of Africa, but did not consider that they led to the replacement of existing Asian populations. In his view, the fact that a "Mongoloid cranial morphology" has been continually present in Asia proved this regional continuity, regardless of any contribution of African genes. This reading of the fossil record arose from the "multiregional theory," of which Wu was one of the initial proponents, as explained in chapter 6. It's easy to see how this theory led him to propose the species name *Homo sapiens daliensis* for the Dali human.

Xinzhi Wu, the Chinese paleoanthropologist and then vice director of the IVPP, in 1999, working at his desk, which is cluttered with casts

Our interpretation is different. While Dali, which is over two hundred thousand years old, shares traits inherited from its common ancestor with Neanderthal and archaic Sapiens, it already has three of the most striking cranial features that distinguish Denisova from European fossils of the same period: a broad skull at the top, a prominent supraorbital ridge interrupted in the middle and a flat rather than snout-like face. If we were to apply the rules of the International Commission on Zoological Nomenclature, Dali would be considered the first Denisovan fossil discovered, and the Denisovan species would be called *Homo daliensis*.

A Chronological Puzzle ... Even for the Biggest of Brains

Unfortunately, the other fossils that have been discovered from the Denisovan period are much smaller. Still, there are some that confirm the features predicted by our initial sketch. This is particularly true of "Xujiayao Man," a steppe hunter specializing in the slaughter of horses, who can provide us with an insight into the lifestyle of Denisovan hunters. Fossils of this human were

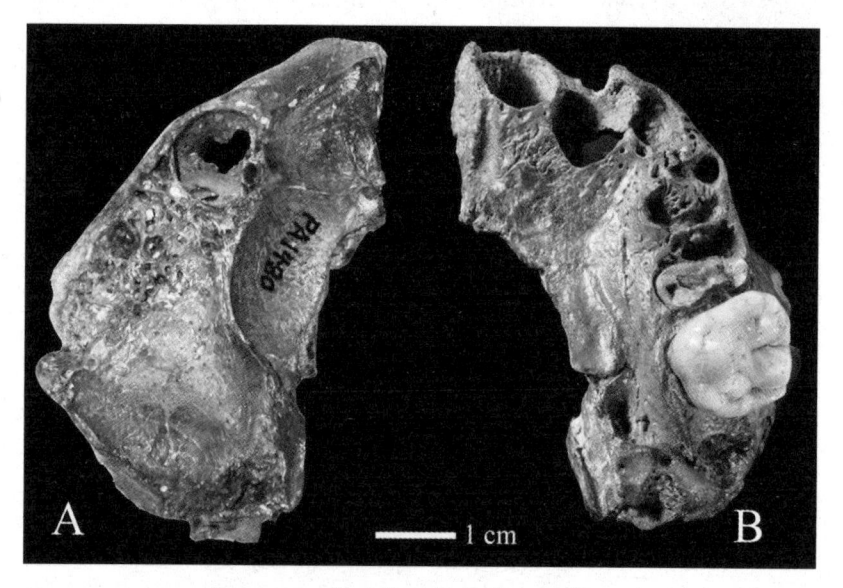

Fragments of maxillary fossils from Xujiayao

unearthed between 1974 and 1979 during IVPP excavations at the Houjiayao and Xujiayao sites, some 60 miles (100 km) west of Beijing, in the Sanggan River Basin. These sites have yielded an incredible abundance of material: more than thirty thousand artifacts and other knapping tools and over five thousand animal remains. The considerable number of cleavers and scrapers reflects an intense level of activity in tasks such as butchery and hide preparation. The human remains discovered at the site include thirteen parietal bones, one temporal bone, two occipital bones, one mandibular bone fragment, one partial juvenile maxilla, and three loose teeth.

This plethora of bones at the open-air sites of Xujiayao and Houjiayao shows that Denisovan hunters butchered the carcasses of the horses they killed on the steppe in large numbers. The butcher's tools found on the site are unsophisticated, consisting mainly of bolas (for crushing bones) and lithic flakes (for cutting and scraping), hastily cut and squared off from stone. If we broaden our geographical lens, a study carried out in 2019 by a team around Shia-Xia Yang along the Sanggan River basin suggests that during the Denisovation period, hunters in the region often produced relatively simple assemblages, comprising a small number of lithic flakes obtained from cores of raw material. This is indicative of opportunistic behavior, especially given that more delicately refined tools have occasionally been found, which these humans were therefore clearly capable of making.

This paints the overall picture that Denisovan hunters traveled extensively across the steppe in pursuit of hordes of horses; after slaughtering several of them, they made use of these horses (their skins, sinews, etc.) presumably before transporting the skins and the best cuts of meat to their living quarters. There, they likely flaked higher quality tools to continue their work. The activities illustrated by the Xujiayao and Houjiayao sites make them similar to Schöningen in Germany (two hundred thousand years old), where horses were also hunted. Unfortunately, they didn't yield the remains of hunting spears of the kind found in Schöningen, but it's very likely that the Denisovan hunters of Xujiayao and Houjiayao also used them, probably after hardening them by fire.

Sadly, it's not clear how old these sites are, despite the numerous endeavors to date them. After no fewer than four attempts and intense discussions between 1984 and 2017 (which dated the sites at 40,000, 50,000, 100,000, and 90,000 to 125,000 years old respectively), a Chinese-Australian team led by Hong Ao, from the Institute of Earth Environment of the Chinese Academy of Sciences in Xi'an, the capital of Shaanxi, resumed the process of trying to date the sites. Combining paleomagnetic data with an estimate obtained through electron spin resonance dating of quartz grains in sediments, these researchers arrived at a time range of 370,000 to 260,000 years. Pending further dating attempts, we can assume that the Xujiayao human lived some 250,000 years ago, in the midst of Denisovation.

Dating in China

It's has always been difficult to date fossils in China. The first reason for this is cultural: Given that prehistoric research was not part of the culture in China for a long time, fossils were instead used as ingredients in traditional medicines and were known as "dragon's bones." This explains why many major fossils have been discovered by chance by farmers who quickly forgot the position of the fossil in the ground, if they remembered where it was found at all. The second reason is the age of the discoveries: Many of the most significant discoveries—such as the Peking Man fossils—were made at the beginning of the twentieth century, when the only possible dating was stratigraphic. Geology also comes into play here, as sedimentation varies according to climate, which in turn varies depending on the era in question. At the Zhoukoudian site, for example, a 164-foot-thick (50 m) sedimentary sequence tells a story spanning at least half a million years. During this period, the climate of the Nihewan Basin often changed, which may have slowed or accelerated the sedimentation process in the cave. When dating sediment, geomorphologists assume that the process of sedimentation followed a steady, uninterrupted pattern over time.

Fortunately, in the second half of the twentieth century, physicists developed so-called direct dating methods. Most of these are radiometric: This involves measuring the residual mass of an isotope (a form of elements that have different numbers of neutrons and different atomic masses) decaying at a regular rate from a starting point, such as the death of an organism. The best-known method is carbon-14 dating, but uranium-thorium dating is also commonly used. A similar principle is cosmogenic isotope dating, using optically stimulated luminescence (OSL) to determine the date on which objects were buried. Unfortunately, these methods come with a number of technical limitations: contamination of samples, displacement of objects over the course of time, complications due to the mode of fossilization, lack of precautions during excavation, etc. As most discoveries made in Asia during the twentieth century were made by farmers, construction workers, etc., prehistorians have only been able to date these later, and very often had to work with fossils that had already long been removed from their context.

As a result, dates obtained in China through direct dating methods do not correlate well with stratigraphic dates, which causes confusion in the fossil record. In terms of the hunt for Denisovans, it has therefore been necessary to study in great detail the precise dating history of each fossil and cross-reference all available information in order to establish the most plausible age.

This age of the Xujiayao and Houjiayao hunters is the first reason we consider them to be Denisovan. The second is their cranial capacity: In 2022, a team led by Xiu-Jie Wu of the IVPP reconstructed a virtual skull from three cranial specimens of the same young adult. They estimated a 95 percent probability that the endocranial volume was between 1,555 and 1,781 cubic centimeters. These values are very high and are in the upper range of variation of Neanderthal cranial capacity.

A third reason—macrodontia—supports our hypothesis that the Xujiayao human was Denisovan: The maxillary fragment of a juvenile individual displays a very large first permanent molar with seven reliefs, which is typical of Denisova (specifically the Denisovan teeth of Baishiya and Denisova). And if we needed a fourth reason, we could take this statement from a publication by Hong Ao's team: "The revised date, combined with the Neanderthal-like features of the *Homo* fossils from Xujiayao, in particular the typically Denisovan molars, implies a possible resemblance between the Xujiayao hominin and early Denisovans."

Without a doubt, this human's big teeth and big head are a sign of an advanced stage of Denisovation. But when can we expect an even more reliable date to be established?

Denisovans Get Big-Headed

This remains to be seen. But other Chinese fossils (the Lingjing skulls) very clearly confirm that the trend of increasing cranial capacity over time among Denisovans mirrored that of Neanderthals. In 2007, Zhan-Yang Li of the IVPP was completing an excavation project at Lingjing, near the town of Xuchang in Henan province, when he came across tools and, two days later, the fragment of a skull. Following this initial discovery, his team uncovered no fewer than forty-five other cranial bones, which could be reassembled into partial cranial vertexes, within six months. These human remains were found alongside "very beautiful" stone and bone tools and a number of animal remains, including giant cervids, gazelles, and woolly rhinoceroses. A series of dating processes using optically stimulated luminescence places these two fragmentary skulls somewhere between 125,000 and 105,000 years ago—i.e., toward the end of an interglacial period even warmer than the present one, in the middle of the Denisovan era.

In 2017, the team that studied these two skulls, including Erik Trinkaus from Washington University in St. Louis, Missouri, noted the proximity between "Lingjing Man" and Neanderthal. Taking into consideration the particularly warm climate that prevailed at the time of the Lingjing human, the researchers saw a resemblance to Neanderthal in this form, which they described as "a new Eurasian human form or an eastern variant of the Neanderthals." They were suggesting that Neanderthals moved along the steppe highway to Asia, where they formed a new lineage with the locals.

This hypothesis is not unreasonable, since two Neanderthal fossils in the Denisova Cave, Denisova 5 and Denisova 9, are dated at 130,300 and 90,900 years old respectively, while the Denisova-Neanderthal hybrid hominin—Denisova 11—is between 118,100 and 79,300 years old. Were there Neanderthals in China after Altai, and if so, did they shape the Asian phenotype as far as Lingjing, in the heart of present-day China? This seems less plausible than the much more parsimonious explanation for the resemblance between Denisova and Neanderthal: They descend from the same ancestor.

Besides, for most other paleoanthropologists, the Lingjing fossils are considered to be Denisovan. When questioned by journalists, Katerina Harvati from the University of Tübingen in Germany, for example, replied that these fossils "have the combination of traits we would expect based on the analysis of ancient Denisovan DNA, which was closely related to Neanderthals."

The enormous volumes of their elongated skulls (around 1,800 cubic centimeters) corresponds to those of the Xujiayao human, which would make them evolved Denisovans. The shape of their supraorbital ridges is also Denisovan. If we add to this their age and the fact that they were found in the middle of the Denisovan territory in the north, we might be tempted to associate them with another "Denisovan" from the south: Maba.

The Ill-Fated Maba Man

The first southern Denisovan ever discovered is no doubt the "Maba Man." It was in 1958 that farmers discovered the partial remains of this individual's skull in a cave near the village of Maba, located in the middle of a humid tropical strip in Southern China. Dating the site through a stratigraphic study proved difficult, as the fossil lay in a vertical fissure inside the cave, but an age was finally established decades later: between 134,000 and 129,000 years old. The ancient fauna identified in the cave includes the great panda and a stegodon, an imposing Asian elephant.

When studying Maba, Wu Rukang described the specimen as "Neanderthaloid"—i.e., "Neanderthal in form." This conclusion may come as a surprise today, but in the 1950s, *Homo erectus* and *Homo neanderthalensis* were the only known species in Eurasia. Still it's hardly surprising if we think about the common traits between Denisova and Neanderthal. In fact, the Maba skull is one of only four fossils that bear testimony to the presence of Denisovans in the south. Its supraorbital ridge is remarkable; unlike that of Neanderthals, it is less pronounced in the center and seems to overflow to the sides, suggesting a broad face. Its cranial capacity is estimated to be around 1,300 cubic centimeters, which places it within the Denisovan range. In addition to this endocranial volume, Maba's age would place him among fully evolved Denisovans, a rare testimony to the continuity of the Denisovan population, from the jungles of Southeast Asia to wintry Siberia, the Tibetan plateau, and the Altai.

But Maba also shows us that the Denisovans of the south had a different lifestyle to those of the north. While the latter slaughtered horses by the hundreds on the cold, open steppes, those in the south had to hunt in the overcrowded forests, where they did not reign supreme. The fang marks visible on Maba's skull show that he was killed by a big cat (a tiger?), which then dragged him by the head into its den after sinking its fangs in—a well-known behavior among big cats. We have this unfortunate end and the cold temperature of the cave to thank for such an extraordinary prehistoric find of an ancient skull in the middle of a tropical zone.

During his lifetime, Maba Man was also the target of another attack—this time human. In 2011, a team of researchers led by Xiu-Jie Wu of the IVPP found signs of a completely closed fracture on the right side of the skull. Since 90 percent of humans are right-handed, this location implies that the individual was struck from behind by a fellow human, perhaps a member of his own group. Erik Trinkaus, who was part of the team, believes this signifies a tumultuous clan life featuring the use of weapons during fights, but also argues that mutual aid took over in the end. Of course, Maba would never have gone on living until the encounter with the cat without the care provided by those around him.

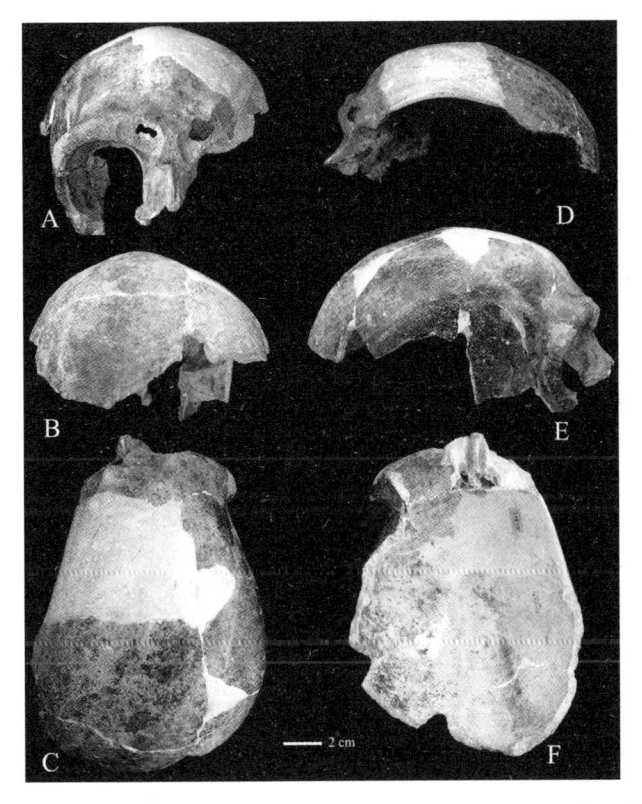

The fossil skull of Maba, reconstructed from available fragments, seen from the front (A), from behind (B), from above (C), from the side (D and E), and from below (F)

A Laotian Denisova

In 2018, a team led by Fabrice Demeter of the French National Museum of Natural History discovered an enormous molar trapped in the hardened sediments of the Cobra Cave, halfway up a limestone cliff in Nam Et-Phou Louey National Park, Laos. Researchers used a variety of methods to date the sedimentary strata containing the tooth, which indicated a range in the middle of the Denisovan era: 164,000 to 131,000 years ago.

Given that DNA conservation is extremely rare in the tropics, the team went straight to studying the series of proteins still contained in the enamel. The proteome was too degraded and therefore could not be used to determine the species, but it did reveal that the tooth had belonged to a woman. Fortunately, morphometric analysis revealed a number of subtle characteristics—enamel thickness, shape of grooves and cusps, dimensions, etc.—which unquestionably place the Cobra Cave tooth in the same category as the Xiahe and Penghu molars. Clearly, Denisova also lived on the Indochinese peninsula.

Penghu, the Denisovan Under the Sea

The next Denisova fossil in our lineup has an astonishing story. In 2008, in part of the Taiwan Strait, fishermen dredged up a half-mandible with four teeth attached, which they sent to Chun-Hsiang Chang of Taiwan's National Museum of Natural Sciences. The team he put together to study the specimen reconstructed the evolution of the sea level in the strait over time, and estimated a maximum age from the depth at which the fossil lay (between 200 and 400 feet/60 and 120 m) of 190,000 years old. Rather imprecise . . .

Despite the usual difficulties encountered in Asia, both on land and under the sea, the clues undoubtedly point to a fossil

Penghu mandible fossil

from the Denisovan period. Analysis of protein sequences carried out by the team of Frido Welker from the University of Copenhagen in 2025 confirmed that the mandible represented that of a male Denisovan. Wear on the two molars prevents us from counting the number of reliefs, but their large size corresponds to the Denisovans. The chin foramen—the hole in the mandibular bone through which nerves and blood vessels pass—is located beneath the premolar, exactly the same spot as on the 160,000-year-old Xiahe mandible, which Penghu closely resembles. This comparison with the only recognized Denisovan mandible suggests that Penghu lived between 200,000 and 100,000 years ago. Remarkably, it also proves that Denisovans living both at high altitude (Xiahe) and on a continental plain now under the sea (Penghu) had exactly the same kind of robust jaw (Penghu was a southern Denisovan).

Denisovans Go Solo

As we have seen several times, fossilization in tropical environments is extremely rare. That's why the *Homo erectus* fossils from the Sangiran site in Java are such a remarkable exception. To what do we owe this special case? Probably the fact that prehistoric people left their dead in the open air, and that in these regions, they mostly set up camp near waterways rather than in the depths of the forest. As a result, the dead—perhaps abandoned after being consumed—were quickly buried by river sediments. Whatever the case, between 1931 and 1933, a team including Gustav von Koenigswald unearthed twelve skulls on one of the banks of the Solo River, not far from the Javanese towns of Ngandong and Sambungmacan. After much debate—as we have seen is often the case in Asia—a widely accepted date was recently attributed to them of between 117,000 and 108,000 years old. Yet in the 1930s, they had been attributed to the *Homo erectus* of Southeast Asia—i.e., associated with the fossils of the Java Man and Peking Man.

Then, in the 1960s, lengthy discussions led to a different classification for them. Paleoanthropologists expressed this view by proposing the species *Homo soloensis*, considered "more evolved" than Java Man. This is putting it mildly, since the two are separated by more than a million years. Then, in the 1980s, these skulls, with cranial capacities far superior to those of the Java Man (over 1,100 cubic centimeters) were classified by some under the *Homo sapiens soloensis* subspecies; they were even claimed this to be the ancestor of the Australian Aborigines.

How plausible is this theory? There are possible traces of early *Homo sapiens* in southern mainland Asia around one hundred thousand years ago, so it is conceivable. However, the theory is not supported by our prior reading of the features of these very robust skulls, which would put them closer to the Harbin skull mentioned in chapter 14.

What's more, the distinctive features of the Solo humans were in fact apparent from the outset—since as early as 1932—as Willem Frederik Florus Oppenoorth (part of the team that found the skulls) wrote in *Scientific American*: "According to his measurements, *Homo soloensis*, as I have named this man, may represent a stage of human development equivalent to that of the Neanderthal 'race.'"

We only need look at the extremely robust supraorbital ridge, which seems to protrude outward from the skull, the extremely thick cranial walls, and the overall appearance of these humans to see their resemblance to Neanderthal.

For us, this is hardly surprising, since the most parsimonious interpretation of Solo skulls is that they are Denisovan. We bet that this interpretation will be backed up by proteomic analysis. Thus, the fact that they share archaic features with Neanderthals is therefore to be expected. And if we take into account the biogeographical context we've gathered so far, the presence of southern Denisovans in this part of the world at this date is just as logical. This is further confirmed by the study of the genetic heritage of this population in the Sapiens genomes of the region, which suggests that interbreeding with Denisovans continued until twenty-five thousand years ago.

Well before the arrival of the main wave of Sapiens who left Africa from seventy thousand years ago, one hundred thousand years ago, southern Denisovans, like the Neanderthals in Europe, were probably numerous in Southeast Asia. This applies in particular to Java, where, like the *Homo erectus* that preceded them, they lived along rivers and were fossilized in the same way.

If we look at Dali, Xujiayao, Lingjing, Maba, Penghu, and Solo, not to mention the fossils we have listed separately in the text box because of their age (Hexian, Chaoxian, Dongzhi), one thing becomes clear: Denisovans lived not only on the Tibetan plateau and in the Altai, but all across Asia. The biogeographical logic of human expansion in Eurasia (see the Earth system, chapter 7) and the distribution of Denisovan DNA in present-day Sapiens populations predicted as much. But before talking about the implications of what we've discovered, it's time to reveal the body and face of Denisova.

The Harbin human, whose skull was discovered under
the most incredible circumstances

14

Fossils Regained:
Completing the Portrait

Before puberty and after menopause, a woman's pelvis tends to be similar to that of a man. That's because its main function is to facilitate walking. During adolescence, estrogen concentration increases in girls, which modulates the shape of the female pelvis, widening it considerably. "In women, the pelvis must be large enough to allow the baby's head to pass through the uterine canal, which is what helps determine whether a specimen is female," explains paleoanthropologist Karen Rosenberg, from the University of Delaware. Well known for her work on the pelvis of Neanderthals, Rosenberg has shown that the pelvis of "Jinniushan Man" was in fact that of a woman in her twenties. And it's the only specimen of a Denisovan body we have.

So far, we've only been able to observe Denisovation in progress from skull specimens. This is why the Jinniushan skeleton is particularly precious, as it comprises large parts and not, as with the Denisova Cave fossil, a mere fingertip. The Jinniushan fossils come from a semi-collapsed limestone cave near the village of Sitian, not far from the town of Dashiqiao in Liaoning province. In addition to human remains, Lü Zuné's team from Peking University, who excavated the cave in 1984, discovered several traces of fire, stone tools, and numerous animal bones. The species found in the cave—beavers, Merck's rhinoceroses, sika deer, saber-toothed felines, etc.—testify to a temperate, even cold climate, which is hardly surprising given that the site is to the northeast of Beijing, at the latitude of North Korea.

Based on the fauna and stratigraphy, Lü Zuné's team first estimated an age range of 310,000 to 160,000 years. Later, Karen Rosenberg, Lü Zuné, and Christopher Ruff of Johns Hopkins University used uranium-thorium dating and electron spin resonance dating, which gave the Jinniushan woman an age of 260,000 years old. This chronological vagueness is typical of Chinese prehistory (see the box on page 000), but the proposed estimate suggests that Jinniushan was alive almost around the same time as the Dali Man and therefore lived during the Denisovan era.

The non-cranial skeletal remains consist of a left ulna (forearm), six vertebrae, ribs, numerous hand and foot bones articulating with each other, a complete left patella, and a left half pelvis (coxal bone and pubis). There is nothing to suggest these bones belonged to more than one individual, as they were found over an area of barely 6.5 square feet (2 sq m); see the following diagram. Then, by an incredible stroke of luck, the Jinniushan skull was also discovered. There was just one problem: An unfortunate encounter with a pickax left the specimen shattered

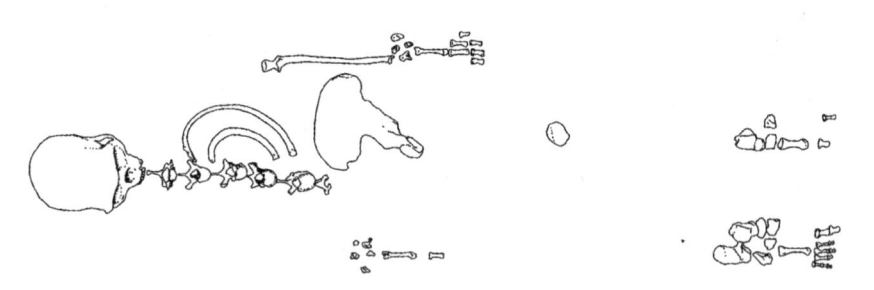

Fragments of the Jinniushan woman skeleton

in one hundred pieces in the ground. The resulting jigsaw puzzle of bones had to be put back together, which is problematic given that the assembly of a fossil is influenced by the prevailing perspective of evolution where it is found. In 1988, Wu Rukang of the IVPP (then one of China's leading paleoanthropologists) and his assistant were the first to attempt a reconstruction. Their attempts were not enough to convince their colleagues at Peking University, however, so the operation was repeated. After much debate, the cranial vault and face were reassembled, but not without bone loss.

A Woman from the North

What does her anatomy tell us? The Jinniushan skull possesses the expected Denisovan features. For a start, it shares archaic features with previously mentioned Denisovan fossils, though it is considerably less robust. For example, the superciliary arches are less prominent than those of Dali, and Jinniushan's skull narrows more after the supraorbital torus. The mastoid apophysis—a conical protrusion at the bottom of the temporal bone—is also smaller than Dali's. His face is broader and flatter, traits already observed in many ancient Asian humans. In short, Jinniushan seems slenderer than her predecessors. Is this because she's a young woman, or because she's "more evolved"? We don't know. But the second possibility suggests it would be worth redating the skeleton.

The Jinniushan woman also has several traits that clearly distinguish her from Dali. Her cranial capacity was between 1,330 and 1,400 cubic centimeters, far greater than Dali's 1,120 cubic centimeters. The cranial walls are also thinner, which could be a feminine trait. Secondly, the fact that the occipital plane and the nape of the neck are at a less acute angle than Dali's suggests a rounding off of the back of the skull, an eminently modern trait. Another derived trait is the widening of the distance between the parietals, a Denisovan trait also observed in Dali and our forensic reconstruction.

The Jinniushan skeleton included an almost complete maxilla and teeth, which is extremely rare. It features the elongated U-shaped dental arch characteristic of Denisovans, another trait indicated by the methylation profile. In Jinniushan, we would also expect to see the characteristic Denisovan macrodontia, but this is absent. In 2000, Jianing He of Peking University studied the teeth and found them to be small compared to those of *Homo erectus*, but also larger than those of modern humans. This could be a sign of evolution, but as we've already seen, Jinniushan's pelvis leaves no room for doubt that she was female. Her solid, wide stature is one of the features highlighted in our forensic reconstruction. The researchers applied biometric laws to estimate Jinniushan's stature and weight, arriving at a height of 5.5 feet (1.68 m) for a weight of 172 pounds (78 kg), which is enormous. Overall, Jinniushan's body proportions are in line with those of a Neanderthal or even an Inuit woman. They reflect a very strong adaptation to the cold.

Yet one fascinating observation seems to illustrate that this adaptation was also cultural: the extreme wear of Jinniushan's teeth, comparable with that of Inuit women in the early twentieth century. It suggests that Jinniushan practiced an activity that was also practiced among Inuit women: skin chewing. This wear is also reminiscent of the state of the teeth of many Neanderthals, who commonly used their teeth as tools; prehistorians refer to this as the "third hand." Admittedly, it seems unwise to simply assume that Denisovan women chewed skins, but since this was also the case with Inuit women, it's plausible that Neanderthals and Denisovans of cold or glacial periods would have done the same.

Harbin Man, the Man Who Came In from the Cold

The Harbin Man is just as remarkable as Jinniushan in more ways than one: This Denisovan skull escaped Japanese occupation in exceptional condition. Sarah Freidline, from the University of Central Florida, best expressed the shock felt by the prehistorian community when she discovered this fossil and its incredible story: "The very complete preservation of the Harbin skull is a dream for any paleoanthropologist."

What's the story behind this dream come true? The specimen came from the outskirts of Harbin, an industrial city in Northern China known for more than one reason: A refuge for White Russian émigrés in the early twentieth century, it was part of Manchukuo, the puppet state created by the Japanese imperial army to carve out a piece of China as a continuation of Korea, which was also annexed. The town of Harbin is also infamous for the construction in 1935 of Unit 731, a center of Japan's bacteriological research system, which used two hundred thousand to three hundred thousand prisoners, or more, as human guinea pigs. Raided by Japanese military police, these Manchu, Chinese, and Korean civilians, including women and children, endured inhumane medical experiments aimed at developing new ways of waging war and depopulating China.

The threat posed by the Japanese explains why, in 1933, a team of Chinese workers building a bridge over the Songhua River carefully avoided reporting that they had found a skull. They did, however, show it to their foreman, who, as luck would have it, was aware of the discovery of the Peking Man in 1929 in Zhoukoudian. Immediately realizing the huge significance of this fossil for his homeland, he hid it in an abandoned well, the traditional method of concealing treasures in China. Once the Sino-Japanese war came to an end, admitting to having worked for the Japanese would mean incrimination. The foreman turned the page and became a farmer, only revealing the hiding place to his grandson on his deathbed in 2018.

The information reached paleoanthropologist Qiang Ji from the China University of Geosciences in Wuhan, who persuaded the family to anonymously donate the fossil to the university museum. And so, Asia's most remarkable prehistoric skull survived the Japanese occupation, the Chinese Civil War, Communism, the Cultural Revolution, and even the devastating trade of "dragon bones"—the informal term for fossils in China—to become, in 2022, the "Dragon Man" himself in a series of events we'll revisit presently. But first, let's take a look at some of the extraordinary fossil's traits.

A Robust Male

The Harbin fossil is crucial, as it confirms and affirms the features observed in other Denisovan skulls. The skull and face were virtually intact and therefore required no reconstruction, making it possible to analyze the specimen with exceptional precision. Unfortunately, however, the mandible is missing, as are all the maxillary teeth except for one invaluable left second molar. Qiang Ji's team has determined that the fossil comes from the upper part of the Huangshan rock formation, which was formed between 309,000 and 138,000 years ago. As for the age of the fossil itself, the most reliable indication we have comes from direct uranium-thorium dating of the strata the skull is thought to come from, which would make it 150,000 years old. This places the Harbin fossil in the middle of the Denisovan era.

Once again, its features are divided between both archaic and derived traits shared with Neanderthal, and derived traits specific to Denisova. The wide orbital and nasal cavities, the great thickness of the bone and the remarkable robustness of the cranial reliefs, particularly the supraorbital ridge that becomes less prominent in the center, are clearly archaic features inherited from *Homo heidelbergensis*. Then there's the rearward elongation of the skull, a low forehead and parietal (temple) bone, and a large cranial capacity of 1,420 cubic centimeters—all features Neanderthal and Denisova have in common. Finally, the exclusively Denisovan features of the Harbin skull are its elevated cranial width (distance between parietals), its enlarged cranial base (distance between temporals) and a much lower face than that of Neanderthal, made very wide by the front cheekbones (zygomatic bones). The orbits are square, which appears to be an accentuation of a trait observed in Dali's skull, in the partially preserved orbit of the Maba skull, as well as in the reconstruction of the Jinniushan woman's face.

Another exclusively Denisovan feature is the elongated, U-shaped dental arch. "The single molar is huge compared to those of Neanderthals," says Qiang Ji. It's also larger than those of the Jinniushan woman. All these cranial and dental features are particularly indicative of Denisova according to the methylation profile study, and have all already been found on other Denisovan fossils: those from Denisova, Xiahe, Dali, Xujiayao, Lingjing, Maba, and Jinniushan.

Qiang Ji's team consider Harbin to be a robust male when compared to the female Jinniushan. Jinniushan has a cranial capacity (1,390 cubic centimeters) similar to Harbin's (1,420 cubic centimeters), but is more gracile, a difference in size that is also reflected in the fossils' respective molars. Harbin's anterior maxillary region is also proportionally wider, its orbits larger and squarer, and its supraorbital ridges thicker. The obvious conclusion is that sexual dimorphism existed in the Denisovan population, as it did in the Neanderthal and archaic Sapiens populations. In short, Jinniushan is a Denisovan woman and Harbin a Denisovan man.

Dragon Man

This is our point of view, which happens to be the predominant one among prehistorians today. The team of paleoanthropologists Qiang Ji and Xijun Ni from the Chinese Academy of Sciences has long been uncertain about the matter. In a press release by his university on September 13, 2018, Qiang Ji first spoke of "China's first fossil of the *Homo heidelbergensis* variety."

He then assembled a team to study the skull, including paleoanthropologist Chris Stringer from the British Natural History Museum. Then, in 2022, when these researchers published their findings, they made the curious assertion that the Harbin human belonged to a new species. To reach this conclusion, Qiang Ji and Xijun Ni applied the statistical method used in prehistory to produce "more or less probable kinship trees" from a selection of traits. After selecting these traits, they arrived at the "most probable kinship tree," which led them to conclude that the Harbin population constituted a "sister group" to *Homo sapiens*.

To name this species, they proposed the binomial *Homo longi*, derived from Longjiang (the "Dragon River"), the name of Heilongjiang province, of which Harbin is the capital.

In an interview with *Pour la Science*, paleoanthropologist Jean-Jacques Hublin of the Collège de France firmly rejected this conclusion: "These researchers have come up with a tree and dates of divergence between its branches that contradict everything else we know today. . . . I find it very hard to understand all these efforts to circumvent the obvious conclusion: All these Chinese fossils are in fact Denisovan, and . . . Denisovans are the sister group of Neanderthals, not *Homo sapiens*."

Since 2022 and the haphazard interpretations of Quiang Ji and Xijun Ni, things have actually gotten . . . worse. After this book was originally published in France, Christopher Bae, from the University of Hawaii, and Xiujie Wu, from the Institute of Vertebrate Paleontology and Paleoanthropology, after studying the Xujiayo-Houjiayao fossils, at the end of 2024 proposed grouping together many of the ancient Asian fossils dating back less than two hundred thousand years. They suggested introducing into the *Homo* genus the "Juluren"—i.e., the new species *Homo juluensis*, or "big-headed human" in Chinese. Their approach is all the more bizarre given that a large head is a trait shared by Sapiens, Neanderthal, and Denisova, meaning it is clearly inherited from their common ancestor *Homo heidelbergensis/rhodesiensis*. It does not, therefore, constitute a criterion for classifying the so-called Juluren as a form in its own right. The two researchers are picking and choosing ancient Chinese fossils, being careful to render to Caesar—or Quiang Ji and Xijun Ni in this case—the things that are Caesar's. In other words, they want to give credit to their colleagues for their already famous find.

They therefore propose grouping together in a single species not only the Xujiayo-Houjiayao fossils, but also those from Lingjing, as well as the already accepted Denisovan fossils: the Xiahe fossils, those from the Denisova cave and the Denisovan tooth from the Laotian Cobra cave, and finally the Penghu mandible discovered at sea off Taiwan. The "Dragon Man," or *Homo longi*, on the other hand, although dating from the same period, has been carefully ignored (even though its only tooth is compellingly reminiscent of the Xiahe and Penghu mandibles), while the Jinniushan woman and the Dali fossil have been added to the mix. In short, the classification of Juluren, the "big-headed man," is difficult to make heads or tails of, and has done nothing more than thicken the scientific fog in China.

In reality, and this is precisely what our investigation demonstrates, given the biogeographical, archaeological, genetic, and paleontological observations we have assembled and organized into a demonstrative whole, Denisova can be considered a paleontological species. This human is now known not only by its genome, but also by the bone and dental features found on fossils from Denisova, Xiahe, Xujiayao, Dali, Lingjing, Maba, Penghu, Jinniushan, and Harbin. This paleontological definition is far from arbitrary, but it is not yet official. In the meantime, most paleoanthropologists agree that there is a lineage in China that is close to, but distinct from, that of Neanderthal: the Denisovan lineage.

From Identikit Portrait to Full Portrait

There's a simple "equation" that can provide us with the most complete portrait possible of Denisova:

Denisova = Harbin skull + Xiahe mandible + Jinniushan body

In other words, the Denisovan mandible is solid, chinless, and endowed with very large teeth; its face is also solid, broad, and fairly flat; the cheekbones (zygomatic bones) are "protruding," quite different from the receding cheekbones of Neanderthals; the orbits are large and rectangular, bordered at the top by a powerful supraorbital ridge, which would be "visored" as in Neanderthals were there not a recess in the center; the body of the maxilla has a hollow above the canine on either side of the nose ("canine fossa"), a feature absent in Neanderthals, which brings Denisova closer to Sapiens in appearance.

The solid appearance of the face when viewed from the front is reinforced by the width of the skull. Behind the face, a large brain is protected by a receding forehead followed by a low, elongated arch, which is very reminiscent of the Neanderthal skull, right down to the nape of the neck.

Moreover, the body of the Jinniushan woman tells us that this large head is complemented by a powerful, stocky body, with a substantial pelvis, short forearms in proportion to the arm (nothing can be said about the proportion of the leg to the thigh), similar to Neanderthals' bodies, which were adapted to cold conditions. This conclusion, however, applies to northern Denisovans but not tropical Denisovans, in the same way it applies only to northern Sapiens, such as the Inuit. Although we don't have a single non-cranial fossil to confirm this, comparison with tropical *Homo sapiens* suggests that southern Denisovans were less stocky and had long, slender limbs that were better able to dissipate heat. This is unless forest Denisovans resembled Pygmies, as the case of the Negritos of the Philippines for example suggests.

It's also worth noting that *Denisova 3*—the only Denisovan for which we have complete, high fidelity sequencing results—possessed genes that, in present-day *Homo sapiens*, are associated with dark skin, brown hair, and brown eyes. This suggests that contact between northern and southern Denisovans led to northern Denisovans retaining the darker features of their tropical counterparts. This Denisovan phenotype contrasts with that of Neanderthals, whose hair ranged from red to chestnut, skins were fair, and eye color varied from brown to blue.

Since both Neanderthals and Denisovans display evidence of

early interbreeding with "super-archaics" (possibly *Homo erectus*), it's likely that this also created a difference in appearance between Denisovans in the south and the north. We don't know how exactly this would have manifested. We can conclude this portrait by noting that, even if we have reconstructed this human from northern fossils, the high level of similarity between the Xiahe and Penghu mandibles, or the resemblance between the molars from the Denisova and Cobra Cave in Laos, acts as a reminder that Denisova, like Asian *Homo sapiens*, belonged to a single species, as confirmed by its genetic heritage.

At this point, we'd like to thank you for following us through these intricate but fascinating considerations on the bone morphology of Denisovans. So far, our portrait constitutes more bones than flesh, so now it's time to go further and bring the Denisovans fully back to life. And there's no doubt that there was a great level of variation across the vast territory of Asia with its diverse climates.

Did southern Denisovan women leave their clans to
form families, like Neanderthals?

15

Bringing Denisova to Life

We've now come to the end of an epic journey. We've explored "generalist" human forms, their various departures from Africa, their evolution into regional forms, the steppe highway and the southern shore of Asia, the Wallace Line, the impact of a mega meteorite, material cultures lost in bamboo forests, and so on. This long road took us from Africa to the Denisova Cave, then farther into prehistoric Asia, whose enigmatic inhabitants we gradually came to understand as part of a vast population that, having once populated the whole of the Far East, left a strong genetic imprint on the current inhabitants of this immense part of the world.

Our goal was to identify the paleontological species of the Denisova Cave dwellers somewhere in Asia, and we've now achieved it with the following formula:

Denisova = Harbin skull + Xiahe mandible + Jinniushan body

We've shown you how to recognize Denisovans in the Asian fossil record, those Prehistoric people who for others are *Homo erectus sensu lato* or *stricto*, evolved *Homo erectus*, *Homo erectus* sapiens, *Homo sapiens daliensis*, *Homo longi*, etc.

We've also developed a unified method for understanding the prehistory of the whole of Eurasia. This paradigm of "generalist human forms" spreading from Africa throughout the history of time is the most prudent way of integrating Eurasian prehistoric data into the greater history of human expansion. This expansion happened time and again in response to changing climates and ecosystems.

The history of every wave of humans leaving Africa, with all their similarities and differences, supports this theory. The fourth wave—Sapiens—is the one we understand best, because it's the most recent. Although we haven't described it in detail here, we have seen how it confirms the relevance of this paradigm. Sapiens has helped us to piece together the scant traces of the third wave of humans from Africa—*Homo heidelbergensis*—and to show that it led simultaneously to Neanderthalization and Denisovation. In Europe, the development of the third generalist human form produced Neanderthal, and no one among paleoanthropologists disputes this. In Asia, this same process produced Denisova, not a "Dragon Man" nor an archaic *Homo sapiens*, that found itself in the Far East of three hundred thousand years ago.

Clearing Up the Muddle in the Middle

Thanks to this new paradigm, we have finally begun to put some semblance of order to the muddle in the middle (see page 138)—i.e., the period of some seven hundred thousand years that begins with the emergence of *Homo heidelbergensis* and ends with that of *Homo sapiens*. And what a relief! According to genetic and archaeological evidence, Neanderthals and Denisovans encountered "super-archaic" humans. The two species gradually became distinct from one another in their respective territories in Asia, moved in different directions, then met and mixed. Some of the details are still unclear: where, when, to what extent, and how many times?

Whatever the case, in western Eurasia, some two hundred thousand years ago, Neanderthal shared material culture and genes with archaic *Homo sapiens*, long before the latter's main departure from Africa (the most significant in terms of genetics) began around seventy thousand years ago. There are many mysteries we still don't have answers to.

It seems evident that ever since humans first tamed fire, the steppe highway has played a previously underestimated role in hot climatic episodes. As a result, the notion of species based on interfertility does not really apply in Eurasia: Neanderthal, Denisova, and Sapiens have all evolved from the same ancestor and have interbred. If you know how to look, the muddle in the middle isn't so muddy at all. Things become clearer if we put the chronology in order, and if we make a clear distinction between the archaic traits of Denisova—common with Neanderthal and archaic Sapiens—and its own traits, which have been observed in fossils that are believed to be Denisovan. In essence, things are simple. European Neanderthals had a sibling who was similar to them in certain ways, but also different: Asian Denisovans.

From Paleontological Portrait to Still Life

Sadly, the portrait we have so far is only a paleontological one. Are there other ways of painting a fuller picture? Yes, but to do so, we first need to better understand the distinction between Asia's two climatic zones. Let's begin with the tropical zone, which runs from the Wallace Line to the Qinling Mountains, at the latitude of Shanghai. We are certain that Denisovans preceded *Homo sapiens* here, since they left an advantageous genetic heritage that facilitated our species's adaptation to the many pathogens of the tropical rainforests in this part of the world.

In the first climatic zone, we can only imagine the life of Denisovan clans hunting small and medium-size animals in the forest labyrinth (tapirs, babirusas—a type of wild boar, snakes, etc.) and occasionally slaughtering large herbivores (buffalo, rhinoceros, elephants) when they managed to wound them by chance or immobilize them in the sinking soil of a swamp. To do this, they probably used wooden or bamboo stakes and assegais (spears) and built traps.

However, the mention of these hunting activities should not blind us to the fact that gathering—generally a collective activity, if the ethnography of today's Sapiens forest societies is anything to go by—probably provided the lion's share of sustenance, whether in the form of grass seeds, tough but edible plants, mushrooms, succulent woodworms, berries, and other fruits. According to anthropologist Alain Testart, the proportion of subsistence derived from gathering is never less than 20 percent (as in the case of the Inuit), and in most cases, particularly in the forest, it is over 60 percent. There's no doubt that this empirical law observed among Sapiens hunter-gatherers also applied to Denisovan hunter-gatherers: For those who know it, the forest is a good provider of resources, whether dietary or medicinal, which can be obtained more easily than prey, and harvested in the relative safety of a group.

In any case, what we know about the Early Paleolithic in Africa and India suggests that these Denisovan clans stayed close to rivers, where there were areas cleared by the grazing of large herbivores or simply by the course of water. They almost certainly liked living in the immense limestone areas of this part of the world and, since they probably tended much more toward lignic than lithic industries, we can only assume that they were also capable of building wood and bamboo dwellings to protect themselves from the dangerous predators that roved about the forest at night: for example, huts surrounded by some kind of barrier or houses raised on pillars.

We don't know how the lives of Denisovans of the southern zones looked, but traditional Papuan houses illustrate that when necessity dictates, humans develop ingenious and diverse solutions to adapt to local conditions. Because they used wood as their primary material, they left no trace of their activities. As we know from the fate of Maba, our only Denisovan from continental southern Asia, besides the deep-sea diver of Penghu and the Laotian tooth, it's also clear that the threat of big cats, starting with the tiger, played a major role in their lives. What else do we know? Perhaps early in their history, tropical Denisovans were likely to have rubbed shoulders with *Gigantopithecus*, enormous cousins of the orangutan that stood 6 to 10 feet (2 to 3 m) tall and weighed between 450 and 1,100 pounds (200 and 500 kg). Like gorillas, these gigantic herbivorous primates (perhaps bamboo eaters like pandas) were probably of little danger to Denisovans, as long as the latter weren't bothering them with any monkey business.

It's worth pointing out once more that all the behaviors we attribute to Asian forest Denisovans is by projection from what we observe among forest-dwelling Sapiens hunter-gatherers. Their way of life is probably close to that of the Denisovans that once lived in the vast area spanning equatorial Asia to tropical continental Asia, as far as the Qinling Mountains. There, the changes in the Earth's ocean undoubtedly split them into isolated groups on several occasions, isolated on large islands and only periodically resuming contact, before finally being joined by *Homo sapiens*.

Steppe to the North

When it comes to understanding the other zone—temperate Asia—we can apply a simple method: comparison with the West. We know that Neanderthals hunted large quantities of meat in one go, which they were able to preserve for certain lengths of time. The evidence of butchery in Xujiayao and Houjiayao, and more generally the sites dating from the Denisovation period in the Sanggan River Basin, illustrate opportunistic behavior comparable to that of the Neanderthals: From pebbles found nearby, tools were hastily flaked to cut up large numbers of carcasses, before transporting the preservable pieces elsewhere.

We also know that, curiously, Neanderthals did not shy away from hunting dangerous animals, such as elephants and aurochs, or even wolves and bears. This is particularly evident at the Biache-Saint-Vaast site in Northern France, where an incredible collection of mammal bones has been found, suggesting that Neanderthals returned to the same hunting station for thousands of years.

Here, the most common species among the 220,000 faunal remains are auroch (69 percent), bear (15.8 percent), and prairie rhinoceros (7.5 percent), though even more dangerous animals such as elephants are also present. However, large, giant, and dangerous species have also been found on Denisovan sites, such as the Merck rhinoceros, giant deer, and saber-toothed felines in the collapsed cave that yielded the Jinniushan Denisovan, or the two elephant species, giant deer, rhinoceros, and horses on the Dali site. In short, like the Neanderthals, it seems that the northern Denisovans had no qualms about confronting large animals.

To kill these animals, Neanderthals used wooden stakes, such as those found at Schöningen and Lehringen. We can imagine that the Denisovans also knew how to select spear shafts, sharpen them, and harden them using fire in order to kill effectively. However, unlike Neanderthals, Denisovans had access to bamboo, a raw material that could be transformed into tools. This no doubt provided them with knives and weapons better suited to hunting small animals.

We also know from isotopic and genetic studies that Neanderthals were well equipped to digest meat and fats. Geneticists have shown that genes selected in these hypercarnivores favored the intake of very large quantities of meat and fat in one go. Variants of these genes are also present in Denisova and are found in Inuit populations and Native Americans (who are descendants of ancient Siberians). This suggests that Denisovans had a fat metabolism similar to Neanderthals, and therefore must have hunted large prey. As with Neanderthals, it's likely that the ability to feed the whole clan at once played a social role for Denisovans.

Moreover, their methylation profile indicates that they possessed large, bell-shaped thoracic cages. Common in Neanderthal skeletons, this torso shape could be linked to the considerable prominence of the liver and kidneys—i.e., the organ responsible for metabolizing excess protein into energy and those specialized in evacuating the nitrogenous products of meat metabolism, respectively. In other words, like their western siblings, Denisovans probably hunted big and had to be able to digest big. No doubt they had high energy requirements in the cold and temperate climate of the northern plains of Asia, not to mention the Altai or the icy Tibetan plateau.

These possibilities, which suggest a strong attachment to meat, do not mean Denisovans in both the north and the south were not also foragers. We know that, depending on the environment in which they lived, Neanderthals used digging sticks to extract tubers from the soil, but also that they ate dates, seeds, shellfish, fish, turtles, water lily heads, not forgetting insects, eggs, honey, snails, frogs, mushrooms, etc. Judging by the diversity of ingredients in Chinese cuisine, we might also presume that in prehistoric Asia, swallow nests, crabs, snakes, and the like all potentially featured.

We have also deduced from the Neanderthals' interest in fur-bearing animals—wolves, bears, foxes, etc.—that they are likely to have clothed themselves in fur. This is supported by the wear on their front teeth, which suggests these humans chewed skins. The condition of the Jinniushan woman's teeth suggests that the same probably applies in eastern Eurasia. Otherwise, how would Denisovans and Neanderthals been able to live and meet one another in the cold of winter in the Altai mountains?

The parallels between west and east indicate that many aspects of Neanderthal life, which have been studied in detail in Europe, can now be studied in Asia, particularly in Northern China, Mongolia, and Korea. There is no doubt that the years to come will bring us more and more studies on the tartar of Denisovan teeth, on the isotopes contained in their bones and teeth, on the genetic traces of the Denisovan presence still inscribed in cave sediments, and so on. The results of this research should confirm the close resemblance between the Neanderthal way of life and that of the northern Denisovans. However, if we understand that Neanderthals remained confined to cold climates by the Mediterranean, in Asia, northern Denisovans were never cut off from the tropical population reservoir conducive to biological evolution and technical innovation.

Why didn't this immense reservoir produce an invasive, generalist species as it did in Africa? There are two crucial geographical factors that can explain. Firstly, the fact that Southeast Asia was alternately a land and an archipelago, while Africa remained a vast network of interconnected habitats. Secondly, the fact that the forest does not seem conducive to the emergence of new forms because it hinders movement over great distances. Even in Africa, the traces left by generalist human forms are numerous everywhere except for the forest. Otherwise, this phenomenon is largely still a mystery. But the fact remains that only Africa has produced generalist human forms one after the other, whose departures in waves changed the course of prehistory in Asia.

Neanderthal and Denisova Live On

In this sense, Neanderthal and Denisova are in a way the last "animal-humans" of Eurasia, who, like other species, lived in balance with nature without ever exhausting it, taking from it only what they needed. They have not completely disappeared and are still with us, since all Eurasians carry between 1.8 and 2.6 percent Neanderthal DNA, while all East Asians carry between 1 and 5 percent Denisovan DNA.

This distribution of Neanderthal and Denisovan genes across Eurasia is further proof of the phenomenon of human waves from Africa. Entering the great continent from Africa, *Homo sapiens* first interbred with Neanderthals in the Near East. Then, those—and only those—who made it to Asia also mixed with Denisovans. Just as Neanderthal is one of the ancestors of today's Europeans, so Denisova is one of the ancestors of modern-day Asians. Denisova, too, lives on inside us.

Further Reading

1. A Secret Comes to Light

Briggs, A. W., et al., "Targeted retrieval and analysis of five Neandertal mtDNA genomes," *Science* 325, no. 5,938 (2009): 318–21.

Condemi, S., and Savatier, F., *A Pocket History of Human Evolution* (The Experiment, 2019).

Green, R. E., et al., "A complete neandertal mitochondrial genome sequence determined by high-throughput sequencing," *Cell* 134, no. 3 (2008): 416–26.

Green, R. E., et al., "A draft sequence of the neandertal genome," *Science* 328, no. 5,979 (2010): 710–22.

Krause, J., et al., "The complete mitochondrial DNA genome of an unknown hominin from southern Siberia," *Nature* 464, no. 7,290 (2010): 894–97.

Krings, M., et al., "Neandertal DNA sequences and the origin of modern humans," *Cell* 90, no. 1 (1997): 19–30.

Pääbo, S., *Neanderthal Man: In Search of Lost Genomes* (Basic Books, 2014).

Reich, D., et al., "Genetic history of an archaic hominin group from Denisova Cave in Siberia," *Nature* 468, no. 23 (2010): 1,053–60.

2. In the Denisova Cave

Arkhipov, S., "Natural habitat of early man in Siberia," *Sborník Geologichých Věd – Antropozoikum* 23 (1999): 133–40.

Bennett, E. A., et al., "Morphology of the Denisovan phalanx closer to modern humans than to Neanderthals," *Science Advances* 5, no. 9 (2019).

Derevianko, A. P., et al., "The Pleistocene peopling of Siberia. A review of environmental and behavioral aspects," *Bulletin of the Indo-Pacific Prehistory Association* 25 (2007).

Higham, T., and Douka, K., "Faire parler les vieux débris," *Pour la Science*, no. 497, published online on February 27, 2019.

Jacobs, Z., et al., "Timing of archaic hominin occupation of Denisova Cave in southern Siberia," *Nature* 565, no. 7,741 (2019): 594–99.

Morley, M. W., et al., "Hominin and animal activities in the microstratigraphic record from Denisova Cave (Altai Mountains, Russia)," *Scientific Reports* 9, no. 13,785 (2019).

Peyrègne, S., et al., "More than a decade of genetic research on the Denisovans," *Nature Reviews Genetics* 25, no. 2 (2024): 83–103.

Prüfer, K., et al., "The complete genome sequence of a Neanderthal from the Altai Mountains," *Nature* 505, no. 7,481 (2014): 43–49.

Shpakova, E. G., and Derevianko, A. P., "The interpretation of odontological features of Pleistocene human remains from the Altai," *Archaeology, Ethnology and Anthropology of Eurasia* 1, no. 1 (2000): 125–38.

Skov, L., et al., "Genetic insights into the social organization of Neanderthals," *Nature* 610, no. 7,932 (2022): 519–25.

Slon, V., et al., "Neandertal and Denisovan DNA from Pleistocene sediments," *Science* 356, no. 6,338 (2017): 605–8.

Slon, V., et al., "The Genome of the offspring of a Neanderthal mother and a Denisovan father," *Nature* 561, no. 7,721 (2018): 113–16.

3. Denisova: A Human Species?

Al-Jâhiz, *Kitāb al-hayawān*, 847.

Boule, M., *L'homme fossile de La Chapelle-aux-Saints*, Masson, 1911, accessed online at archive.org/details/b22463355.

Bürger, W., "Fuhlrott, Johann Carl," *Deutsche Biographie*, 1961, accessed online at deutsche-biographie.de/gnd119519763. html?language=de#ndbcontent.

Condemi, S., and Savatier, F., *Néandertal, mon frère* (Champs-Flammarion, 2019).

King, W., "The reputed fossil man of the Neanderthal," *The Quarterly Journal of Science* vol. 1 (1864), 88–97, accessed online at biodiversitylibrary.org/part/204441.

Lecointre, G., and Le Guyader, H., *Classification phylogénétique du vivant*, vol. 1, 4th ed. (Belin, 2016).

Linnâe, C., *Systema naturæ*, 6th ed. (Kiesewetter, 1748), accessed online at gdz.sub.uni-goettingen.de/id/PPN371257700.

Mayr, E., *Systematics and the Origin of Species* (Columbia University Press, 1942).

4. The Story of Denisova as Told by Genes

Condemi, S., et al., "Blood groups of Neandertals and Denisova decrypted," *PLOS One* 16, no. 7 (2021).

Darwin, C., *On the Origin of Species by Means of Natural Selection, or the Preservation of Favoured Races in the Struggle for Life* (John Murray, 1859).

Hadjinjak, M., et al., "Reconstructing the genetic history of late Neanderthals," *Nature* 555 (2018): 652–56.

Krause, J., et al., "The complete mitochondrial DNA genome of an unknown hominin from southern Siberia," *Nature* 464 (2010): 894–97.

Matthias, M., et al., "A mitochondrial genome sequence of a hominin from Sima de los Huesos," *Nature* 505 (2013): 403–6.

Petr, M., et al., "The Evolutionary History of Neanderthal and Denisovan Y chromosomes," *Science* 369 (2020): 1,653–56.

Posth, C., et al., "Deeply divergent archaic mitochondrial genome provides lower time boundary for African gene flow into Neanderthals," *Nature Communications* 8 (2017).

Slon, V., et al., "Neandertal and Denisovan DNA from Pleistocene sediments," *Science* 356 (2017): 605–8.

Zhang, D., et al., "Denisovan DNA in Late Pleistocene sediments from Baishiya Karst Cave on the Tibetan Plateau," *Science* 370 (2020): 584–87.

5. A Vast Empire in the East

Betti, L., et al., "Distance from Africa, not climate, explains within population phenotypic diversity in humans," *Proceedings of the Royal Society of London B: Biological Sciences* 276 (2009): 809–14.

Choin, J., et al., "Genomic insights into population history and biological adaptation in Oceania," *Nature* 592 (2021): 583–89.

GenomeAsia100K Consortium, "The GenomeAsia 100K Project enables genetic discoveries across Asia," *Nature* 576 (2019): 106–11.

Jacobs, G. S., et al., "Multiple deeply divergent Denisovan ancestries in Papuans," *Cell* 177, no. 4 (2019): 1,010–21.

Larena, M., et al., "Philippine Ayta possess the highest level of Denisovan ancestry in the world," *Current Biology* 31, no. 19 (2021): 4,219–30.

Mondal, M., et al., "Genomic analysis of Andamanese provides insights into ancient human migration into Asia and adaptation," *Nature Genetics* 48, no. 9 (2016): 1,066–70.

Quintana-Murci, L., *Human Peoples: On the Genetic Traces of Human Evolution, Migration and Adaptation* (Allen Lane, 2025).

Reich, D., et al., "Genetic history of an archaic hominin group from Denisova Cave in Siberia," *Nature* 468, no. 23 (2010): 1,053–60.

Reyes-Centeno, H., et al., "Genomic and cranial phenotype data support multiple modern human dispersals from Africa and a southern route into Asia," *PNAS* 111, no. 20 (2014): 7,248–53.

Skoglund, P., et al., "Genomic insights into the peopling of the Southwest Pacific," *Nature* 538 (2016): 510–13.

Voris, H. K., "Maps of Pleistocene sea levels in southeast Asia: Shorelines, river systems and time durations," *Journal of Biogeography* 27, no. 5 (2000): 1,153–67.

6. Before Denisova: The Curious Case of *Homo erectus*

Antón, S. C., "Natural History of *Homo erectus*," *American Journal of Physical Anthropology* 122, no. S37 (2003): 126–70.

Belmaker, M., et al., "New evidence of hominid presence in the Lower Pleistocene of the Southern Levant," *Journal of Human Evolution* 43, no. 1 (2002): 43–56.

Berger, L. R., et al., "Investigation of a credible report by a US Marine on the location of the missing Peking Man fossils," *South African Journal of Science* 108, no. 3–4 (2012): 6–8.

Bermúdez de Castro, J. M., et al., "Early Pleistocene human mandible from Sima del Elefante (TE) cave site in Sierra de Atapuerca (Spain): A comparative morphological study," *Journal of Human Evolution* 61, no. 1 (2011): 12–25.

Carbonell, E., et al., "The first hominin of Europe," *Nature* 452 (2008): 465–69.

Dennell, R., *The Palaeolithic Settlement of Asia* (Cambridge University Press, 2008), 147.

Eckhardt, R. B., et al., "Multiregional Evolution," *Science* 262 (1993): 973–74.

Haeckel, E., *Natürliche Schöpfungsgeschichte* (Reimer, 1868), accessed online at archive.org/details/natrlichesch1868haec.

Herries, A. I. R., et al., "Contemporaneity of *Australopithecus, Paranthropus*, and early *Homo erectus* in South Africa," *Science* 368 (2020).

von Koenigswald, G. H. R., and Weidenreich, F., "The Relationship between *Pithecanthropus* and *Sinanthropus*," *Nature* 144 (1939): 926–29.

Kubat, J., et al., "Dietary strategies of Pleistocene *Pongo* spp. and *Homo erectus* on Java (Indonesia)," *Nature Ecology & Evolution* 7, no. 2 (2023): 279–89.

Leakey, R. E., "Further Evidence of lower Pleistocene hominids from East Rudolf, North Kenya, 1971," *Nature* 237 (1972): 264–69.

Mussi, M., et al., "Early *Homo erectus* lived at high altitudes and produced both Oldowan and Acheulean tools," *Science*, vol. 382, no. 6,671, (2023): 713–18.

Sautman, B., "Peking Man and the politics of paleoanthropological nationalism in China," *The Journal of Asian Studies*, vol. 60, no. 1 (2001): 95–200.

Shen, G., et al., "Age of Zhoukoudian *Homo erectus* determined with ^{26}Al/^{10}Be burial dating," *Nature* 458 (2009): 198–200.

Shuji, M., et al., "Age control of the first appearance datum for Javanese *Homo erectus* in the Sangiran area," *Science* 367 (2020): 210–14.

Tattersall, I., *Understanding Human Evolution* (Cambridge University Press, 2022), 225.

Walker, A., and Leakey, R., eds., *The Nariokotome* Homo erectus *Skeleton* (Harvard University Press, 1993).

Wayman, E., "Mystery of the lost Peking man fossils solved?," *Smithonian Magazine*, published online on March 28, 2012.

Weidenreich, F., "The Mandibles of *Sinanthropus pekinensis*: A comparative study," *Palaeontologia Sinica*, series D, vol. 7, fascicle 3 (1936).

Weidenreich, F., "The extremity bones of *Sinanthropus pekinensis*," *Palaeontologia Sinica*, new series D, no. 5 (1941).

Weidenreich, F., "The Skull of *Sinanthropus pekinensis*: A comparative study on a primitive hominid skull," *Palaeontologica Sinica*, new series D, no. 10 (1943).

Weidenreich, F., *Apes, Giants, and Man* (University of Chicago Press, 1946), 82.

Wu, X., et al., "Morphological and morphometric analysis of variation in the Zhoukoudian *Homo erectus* brain endocasts," *Quaternary International* 211, no. 1–2 (2010): 4–13.

Wu, X., and Poirier, F. E., *Human Evolution in China: A Metric Description of the Fossils and a Review of the Sites* (Oxford University Press, 1995).

Yamamoto, K., "The myth of 'Nihonjinron,' homogeneity of Japan and its influence on the society," *CERS Working Paper*, 2015, accessed online at cers.leeds.ac.uk/wp-content/uploads/sites/97/2016/04/The-myth-of-%E2%80%9CNihonjinron%E2%80%9D-homogeneity-of-Japan-and-its-influence-on-the-society-Kana-Yamamoto.pdf.

Zaim, Y., et al., "New 1.5 million-year-old *Homo erectus* maxilla from Sangiran (Central Java, Indonesia)," *Journal of Human Evolution* 61, no. 4 (2011): 363–76.

Zanolli, C., et al., "Inner tooth morphology of *Homo erectus* from Zhoukoudian. New evidence from an old collection housed at Uppsala University, Sweden," *Journal of Human Evolution* 116 (2018): 1–13.

Zeitoun, V., et al., "Solo man in question: Convergent views to split Indonesian *Homo erectus* in two categories," *Quaternary International* 223–24 (2010): 281–92.

7. Earth Systems: The Science of Denisova

Biton, E., et al., "Red Sea during the Last Glacial Maximum: Implications for sea level reconstruction," *Paleoceanography* 23, no. 1, 2008.

Boulay, S., *Enregistrements sédimentaires des variations de la mousson sud-est asiatique au cours des 2 derniers millions d'années*, doctoral thesis in mineralogy (Paris-Sud University, 2003).

Collina-Girard, J., "Geology of Gibraltar Strait and the Atlantis myth," *Bulletin de la Societe Vaudoise des Sciences Naturelles* 88, no. 3 (2003): 323–41.

Détroit, F., et al., "A new species of *Homo* from the Late Pleistocene of the Philippines," *Nature*, vol. 568, no. 7,751 (2019): 181–86.

Morwood, M. J., et al., "Archaeology and age of a new hominin from Flores in eastern Indonesia," *Nature*, vol. 431, no. 7,012 (2004): 1,087–91.

Morwood, M. J., et al., "Further evidence for small-bodied hominins from the Late Pleistocene of Flores, Indonesia," *Nature*, vol. 437, no. 7,061 (2005): 1,012–17.

Raikes, R. L., and Dyson, R. H., "The prehistoric climate of the Baluchistan and the Indus valley," *American Anthropologist*, vol. 63, no. 2 (1961): 265–81.

Ramstein, G., *Voyage à travers les climats de la Terre*, Odile Jacob, 2015.

Voisin, J., "*Homo naledi*, le grimpeur qui n'était plus un singe," *Pour la Science*, no. 524, published online on May 21, 2021.

Voris, H. K., "Maps of Pleistocene sea levels in southeast Asia: Shorelines, river systems and time durations," *Journal of Biogeography*, vol. 27, no. 5 (2000): 1,153–67.

Yavuz, V., et al., "The frozen Bosphorus and its paleoclimatic implications based on a summary of the historical data," in *The Black Sea Flood Question: Changes in Coastline, Climate and Human Settlement*, Yanko-Hombach, V., et al., eds. (Springer, 2007), 633–49.

8. The Melting Pot That Made Denisova

Arkhipov, S., "Natural habitat of early man in Siberia," *Sborník Geologichých Věd – Antropozoikum* 23 (1999): 133–40.

Arzarello, M., et al., "The Pirro Nord site (Apricena, Fg, Southern Italy) in the context of the first European peopling: Convergences and divergences," *Quaternary International* 389 (2015): 255–63.

Ashton, N., et al., "Hominin Footprints from Early Pleistocene Deposits at Happisburgh, UK," *PLOS One* 9, no. 2 (2014).

Barash, A., et al., "The earliest Pleistocene record of a large-bodied hominin from the Levant supports two out-of-Africa dispersal events," *Scientific Reports* 12 (2022).

Bourguignon, L., et al., "Bois-de-Riquet (Lézignan-la-Cèbe, Hérault): A late Early Pleistocene archeological occurrence in southern France," *Quaternary International* 393 (2016): 24–40.

Brumm, A., et al., "Stone technology at the Middle Pleistocene site of Mata Menge, Flores, Indonesia," *Journal of Archaeological Science* 37, no. 3 (2010): 451–73.

Carbonell, E., et al., "The first hominin of Europe," *Nature* 452 (2008): 465–69.

Carotenuto, F., et al., "Venturing out safely: The biogeography of *Homo erectus* dispersal out of Africa," *Journal of Human Evolution* 95 (2016): 1–12.

Clark, G., *The Stone Age Hunters* (Thames and Hudson, 1967).

Darwin, C., *The Voyage of the Beagle: A Naturalist's Voyage Round the World* (H. Colbourn, 1839).

Gaillard, C., et al., "Lower and Early Middle Pleistocene Acheulian in the Indian sub-continent," *Quaternary International* 223–224 (2010): 234–41.

Han, F., et al., "The earliest evidence of hominid settlement in China: Combined electron spin resonance and uranium series (ESR/U-series) dating of mammalian fossil teeth from Longgupo cave," *Quaternary International* 434, part A (2017): 75–83.

Li, H., et al., "Longgudong, an Early Pleistocene site in Jianshi, South China, with stratigraphic association of human teeth and lithics," *Science China Earth Sciences* 60, no. 3 (2017): 452–62.

Lordkipanidze, D., et al., "A Complete Skull from Dmanisi, Georgia, and the Evolutionary Biology of Early *Homo*," *Science* 342 (2013): 326–31.

Moreau, C., et al., "Native American Admixture in the Quebec Founder Population," *PLOS One* 8, no. 6 (2013).

Mulvaney, D. J., "The Pajitanian industry: Some observations," *Mankind* 7, no. 3 (1970): 184–87.

Pappu, S., et al., "Early Pleistocene Presence of Acheulian Hominins in South India," *Science* 331 (2011): 1,596–99.

Pappu, S., and Akhilesh, K., "Tools, trails and time: Debating Acheulian group size at Attirampakkam, India," *Journal of Human Evolution* 130 (2019): 109–25.

Pawlik, A., "Is the functional approach helpful to overcome the typology dilemma of lithic archaeology in Southeast Asia?," *Bulletin of the Indo-Pacific Prehistory Association* 29 (2009): 6–14.

Rogers, A. R., et al., "Neanderthal-Denisovan ancestors interbred with a distantly related hominin," *Science Advances* 6, no. 8 (2020).

Ronen, A., "The oldest human groups in the Levant," *Comptes Rendus Palevol* no. 1–2 (2006): 343–51.

Shen, G., et al., "Age of Zhoukoudian *Homo erectus* determined with ^{26}Al/^{10}Be burial dating," *Nature* 458 (2009): 198–200.

Shunkov, M., "The characteristics of the Altai (Russia) Middle Paleolithic in regional context," *Bulletin of the Indo-Pacific Prehistory Association* 25 (2005): 69–77.

Sirakov, N., et al., "An ancient continuous human presence in the Balkans and the beginnings of the settlement of western Eurasia: A Lower Pleistocene example of Lower Palaeolithic in Kozarnika cave (North-western Bulgaria)," *Quaternary International* 223–224 (2010): 94–106.

Toro-Moyano, I., et al., "The oldest human fossil in Europe, from Orce (Spain)," *Journal of Human Evolution* 65, no. 1 (2013): 1–9.

Vialet, A., et al., "Proposition de reconstitution du deuxième crâne d'*Homo erectus* de Yunxian (Chine)," *Comptes Rendus Palevol* 4, no. 3 (2005): 265–74.

Vialet, A., et al., "Homo erectus found still further west: Reconstruction of the Kocabaş cranium (Denizli, Turkey)," *Comptes Rendus Palevol* 11, no. 2–3 (2012): 89–95.

Zhu, Z., et al., "New dating of the *Homo erectus* cranium from Lantian (Gongwangling), China," *Journal of Human Evolution* 78 (2015): 144–57.

9. Denisova and Neanderthal: A Common Ancestor

Adler, D. S., et al., "Early Levallois technology and the Lower to Middle Paleolithic transition in the Southern Caucasus," *Science* 345 (2014): 1,609–13.

Alperson-Afil, N., and Goren-Inbar, N., "Out of Africa and into Eurasia through controlled use of fire: Evidence from Gesher Benot Ya'aqov, Israel," *Archaeology Anthropology & Ethnology of Eurasia* 28, no. 1 (2006): 63–78.

Alperson-Afil, N., et al., "Spatial Organization of Hominin Activities at Gesher Benot Ya'aqov, Israel," *Science* 326 (2009): 1,677–80.

Antoine, P., et al., "Rapport de fouille. Opération Abbeville -Moulin Quignon Avril 2017. Autorisation de fouille programmée DRAC Picardie : n°2017-01," *Ministère de la Culture et de Communication* (2019): 71.

Arambourg, C., "Le gisement de Ternifine et l'Atlanthropus," *Société Préhistorique Française* 52, no. 1–2 (1955): 94–95.

Condemi, S., and Savatier, F., *Néandertal, mon frère* (Champs-Flammarion, 2019).

Daura, J., et al., "New Middle Pleistocene hominin cranium from Gruta da Aroeira (Portugal)," *PNAS* 114, no. 13 (2017): 3,397–402.

Geraads, D., et al., "The Pleistocene hominid site of Ternifine, Algeria: New results on the environment, age, and human industries," *Quaternary Research* 25, no. 3 (1986): 380–86.

Gray, G. M., "Starch Digestion and Absorption in Nonruminants," *The Journal of Nutrition* 122, no. 1 (1992): 172–77.

Grün, R., et al., "Dating the skull from Broken Hill, Zambia, and its position in human evolution," *Nature* 580 (2020): 372–75.

Harmon, D. L., "Review: Nutritional regulation of intestinal starch and protein assimilation in ruminants," *Animal* 14, no. S1 (2020): 17–28.

Herries, A. I. R., "A Chronological Perspective on the Acheulian and Its Transition to the Middle Stone Age in Southern Africa: The Question of the Fauresmith," *International Journal of Evolutionary Biology* (2011).

Heyes, P. J., et al., "Selection and Use of Manganese Dioxide by Neanderthals," *Scientific Reports* 6 (2016).

Johnson, C. R., and McBrearty S., "500,000 year old blades from the Kapthurin Formation, Kenya," *Journal of Human Evolution* 58, no. 2 (2010): 193–200.

Moncel, M., et al., "A biface production older than 600 ka ago at Notarchirico (Southern Italy) contribution to understanding early Acheulean cognition and skills in Europe," *PLOS One* 14, no. 10 (2019).

Savatier, F., "Le plus vieux des Portugais a 400 000 ans," *Pour la Science*, published online on April 8, 2017.

Schmidt, P., "Comment les Néandertaliens fabriquaient du goudron," *Pour la Science*, no. 520, published online on February 5, 2021.

Sorensen, A. C., et al., "Neandertal fire-making technology inferred from microwear analysis," *Scientific Reports* 8 (2018).

Stringer, C. B., et al., "The Middle Pleistocene human tibia from Boxgrove," *Journal of Human Evolution* 34, no. 5 (1998): 509–47.

Thieme, H., "Lower Palaeolithic hunting spears from Germany," *Nature* vol. 385 (1997): 807–10.

Tryon, A. C., et al., "Levallois Lithic Technology from the Kapthurin Formation, Kenya: Acheulian Origin and Middle Stone Age Diversity," *African Archaeological Review* 22, no. 4 (2005): 199–229.

Wrangham, R., *Catching Fire: How Cooking Made Us Human* (Basic Books, 2009).

10. To the East

Antón, S. C., "Natural History of *Homo erectus*," *American Journal of Physical Anthropology* 122, no. S37 (2003): 126–70.

Athreya, S., "Was *Homo heidelbergensis* in South Asia? A test using the Narmada fossil from central India," in *The Evolution and History of Human Populations in South Asia: Inter-disciplinary*

Studies in Archaeology, Biological Anthropology, Linguistics and Genetics, Petraglia, M. D., and Allchin, B., eds. (Springer, 2007), 137–70.

Brumm, A., and Moore, M. W., "Biface distributions and the Movius Line: A Southeast Asian Perspective," *Australian Archaeology* 74, no. 1 (2012): 34–46.

Dennell, R. W., "The Nihewan Basin of North China in the Early Pleistocene: Continuous and flourishing, or discontinuous, infrequent and ephemeral occupation?" *Quaternary International* 295 (2013): 223–36.

Derevianko, O. P., et al., "Who were the Denisovans?" *Archaeology, Ethnology and Anthropology of Eurasia* 48, no. 3 (2020): 3–21.

Gaillard, C., et al., "Technological analysis of the Acheulian assemblage from Atbarapur in the Siwalik Range (Hoshiarpur District, Punjab)," *Man and Environment* 33, no. 2 (2008): 1–14.

Hou, Y., et al., "Mid-Pleistocene Acheulean-like Stone Technology of the Bose Basin, South China," *Science* 287 (2000): 1,622–26.

Hyodo, M., et al., "High-resolution record of the Matuyama-Brunhes transition constrains the age of Javanese *Homo erectus* in the Sangiran dome, Indonesia," *PNAS* 108, no. 49 (2011): 19,563–68.

Kennedy, A. R., et al., "Is the Narmada hominid an Indian *Homo erectus*?" *American Journal of Biological Anthropology* 86, no. 4 (1991): 475–96.

de Lumley, H., et al., *Les industries lithiques du Paléolithique ancien du bassin de Bose* (CNRS Éditions, 2020).

Lumley, M. A., and Sonakia, A., "Première découverte d'un *Homo erectus* sur le continent indien à Hathnora, dans la moyenne vallée de la Narmada," *L'Anthropologie* 89, no. 1 (1985): 13–61.

Lycett, S. J., "Is the Soanian techno-complex a Mode 1 or Mode 3 phenomenon? A morphometric assessment," *Journal of Archaeological Science* 34, no. 9 (2007): 1,434–40.

Pappu, S., et al., "Early Pleistocene Presence of Acheulian Hominins in South India," *Science* 331 (2011): 1,596–99.

Sémah, F., "Le genre *Homo* dans les archipels du sud-est asiatique: Isolement et diversité," in *Catalogue de l'exposition Homo Faber au musée national de préhistoire des Eyzies* (Éditions de la Réunion des musées nationaux, 2021), 118–21.

Sieh, K., et al., "Australasian impact crater buried under the Bolaven volcanic field, Southern Laos," *PNAS* no. 3 (2019): 1,346–53.

Sieh, K., et al., "Proximal ejecta of the Bolaven extraterrestrial impact, southern Laos," *PNAS* 120, no. 50 (2023).

Sonakia, A., and Biswas, S., "Antiquity of the Narmada *Homo erectus*, the early man of India," *Current Science* 75, no. 4 (1998): 391–93.

Widianto, H., et al., "The emergence and distribution of early modern humans in Indonesia," *L'Anthropologie* 127, no. 3 (2023): 10316.

11. The Bamboo Empire

Bar-Yosef, O., "Chinese Paleolithic challenges for interpretations of Paleolithic archaeology," *Anthropologie* 53, no. 1–2 (2015): 77–92.

Bar-Yosef, O., et al., "Were bamboo tools made in prehistoric Southeast Asia? An experimental view from South China," *Quaternary International* 269 (2012): 9–21.

Conard, N. J., et al., "A 300,000-year-old throwing stick from Schöningen, northern Germany, documents the evolution of human hunting," *Nature Ecology & Evolution* 4, no. 5 (2020): 690–93.

Forestier, H., "Des outils nés de la forêt," in *Peuplement anciens et actuels des forêts tropicales*, Froment, A., and Guffroy, J., eds. (IRD Éditions, 2003), 315–37.

Forestier, H., et al., "Reduction Sequences During the Hoabinhian Technocomplex in Cambodia and Thailand: A New Knapping Strategy in Southeast Asia from the Terminal Upper Pleistocene to mid Holocene," *Lithic Technology* 47, no. 2 (2022): 147–70.

Hutson, J. M., et al., "Revised age for Schöningen hunting spears indicates intensification of Neanderthal cooperative behavior around 200,000 years ago," *Science Advances* 11, no. 19 (2025).

Li, G., et al., "Chronology and paleoclimatic context of hominin occupations in the Fenhe River Basin of northern China during the middle to Late Pleistocene," *Quaternary Science Reviews* 326 (2024).

MacDonald, K., et al., "Middle Pleistocene fire use: The first signal of widespread cultural diffusion in human evolution," *PNAS* 118, no. 31 (2021).

Savatier, F., "La grotte de Denisova enfin bien datée," *Pour la Science*, published online on February 15, 2019.

Savatier, F., "De l'ADN dénisovien découvert hors de la grotte de Denisova," *Pour la Science*, published online on December 18, 2020.

Thieme, H., "Die ältesten Speere der Welt – Fundplätze der frühen Altsteinzeit im Tagebau Schöningen," *Archäologisches Nachrichtenblatt* 10 (2005): 409–17.

12. A Forensic Reconstruction

Bailey, S. E., and Liu, W., "A comparative dental metrical and morphological analysis of a Middle Pleistocene hominid maxilla from Chaoxian (Chaohu)," *Quaternary International* 211, no. 1–2 (2010): 14–23.

Chen, F., et al., "A late Middle Pleistocene Denisovan mandible from the Tibetan Plateau," *Nature* 569 (2019): 409–12.

Cui, Y., and Wu, X., "A geometric morphometric study of a Middle Pleistocene cranium from Hexian, China," *Journal of Human Evolution* 88, no. 6 (2015): 54–69.

Dong, W., "Biochronological framework of *Homo erectus* horizons in China," *Quaternary International* 400 (2016): 47–57.

Gokhman, D., et al., "Reconstructing Denisovan Anatomy Using DNA Methylation Maps," *Cell* 179, no. 1 (2019): 180–92.

Grün, R., et al., "ESR and U-series analyses of teeth from the palaeoanthropological site of Hexian, Anhui Province, China," *Journal of Human Evolution* 34, no. 6 (1998): 555–64.

Guanjun, S., et al., "Re-examination of the chronological position of Chaoxian Man," *Acta Anthropologica Sinica* 13, no. 3 (1994): 249–56.

Liu, W., and Wu, X., "The Hominid Fossils from China Contemporaneous with the Neanderthals and Some Related Studies," in *Continuity and Discontinuity in the Peopling of Europe: One Hundred Fifty Years of Neanderthal Study*, Condemi, S., and Weniger, G., eds. (Springer, 2011), 77–87.

Peyrègne, S., et al., "More than a decade of genetic research on the Denisovans," *Nature Reviews Genetics* 25, no. 2 (2024): 83–103.

Savatier, F., "*Homo erectus* était bien un omnivore opportuniste," *Pour la Science*, published online on March 8, 2023.

Sun, X., et al., "TT-OSL and post-IR IRSL dating of the Dali Man site in central China," *Quaternary International* 434, part A (2017): 99–106.

Tiemei, C., et al., "Antiquity of *Homo sapiens* in China," *Nature* 368 (1994): 55–56.

Vicedomini, R., et al., "Genetic dietary adaptation in Neandertal, Denisovan and Sapiens revealed by gene copy number variation," *bioRxiv* (2021).

Wu, R., and Dong, X., "Preliminary study of *Homo erectus* remains from Hexian, Anhui," *Acta Anthropologica Sinica* 1, no. 1 (1982): 2–102.

Wu, X., and Poirier, F. E., *Human Evolution in China: A Metric Description of the Fossils and a Review of the Sites* (Oxford University Press, 1995).

Xia, L., "Chinese researchers discover 300,000-year-old ancient human fossils," *Xinhhua*, published online on May 24, 2019.

Zhang, D., et al., "Denisovan DNA in Late Pleistocene sediments from Baishiya Karst Cave on the Tibetan Plateau," *Science* 370 (2020): 584–87.

13. In Search of Lost Fossils

Ao, H., et al., "An updated age for the Xujiayao hominin from the Nihewan Basin, North China: Implications for Middle Pleistocene human evolution in East Asia," *Journal of Human Evolution* 106 (2017): 54–65.

Bae, C. J., *The Paleoanthropology of Eastern Asia* (University of Hawaii Press, 2024).

Bae, C. J., and Wu, X., "Making sense of eastern Asian Late Quaternary hominin variability," *Nature Communications* 15, no. 9479 (2024).

Chang, C., et al., "The first archaic *Homo* from Taiwan," *Nature Communications* 6 (2015).

Condemi, S., and Savatier, F., "Pourquoi Dénisova ne doit pas devenir *Homo juluensis*," *Pour La Science*, February 25, 2025.

Demeter, F., et al., "A Middle Pleistocene Denisovan molar from the Annamite Chain of northern Laos," *Nature Communications* 13 (2022).

Kaifu, K., and Athreya, S., "Diversity and evolution of archaic eastern Asian hominins: A synthetic model of the fossil and genetic records," *PaleoAnthropology*, 2024.

Li, Z., et al., "Late Pleistocene archaic human crania from Xuchang, China," *Science* 355 (2017): 969–72.

McDonald, F., "Ancient Skulls Found in China Could Belong to an Unknown Human Species," Science Alert, published online on March 4, 2017.

Sawafuji, R., et al., "East and Southeast Asian hominin dispersal and evolution: A review," *Quaternary Science Reviews* 333, no. 108669 (2024).

Tiemei, C., et al., "Antiquity of *Homo sapiens* in China," *Nature* 368 (1994): 55–56.

Tsutaya, T., et al., , "A male Denisovan mandible from Pleistocene Taiwan," *Science*, April 10, 2025.

Tu, H., et al., "[26]Al/[10]Be Burial Dating of Xujiayao-Houjiayao Site in Nihewan Basin, Northern China," *PLOS One* 10, no. 2 (2015).

Woo, J., and Peng, R., "Fossil human skull of early paleoanthropic stage found at Mapa, Shaoquan, Kwangtung Province," *Vertebrata Palasiatica* 3, no. 4 (1959): 176–82.

Wu, X., "A well-preserved cranium of an archaic type of early *Homo sapiens* from Dali, China," *Scientia Sinica* 24, no. 4 (1981): 530–41.

Wu, X., et al., "Antemortem trauma and survival in the late Middle Pleistocene human cranium from Maba, South China," *PNAS* 108, no. 49 (2011): 19,558–62.

Wu, X., et al., "Evolution of cranial capacity revisited: A view from the late Middle Pleistocene cranium from Xujiayao, China," *Journal of Human Evolution* 163 (2022): 103–19.

Xiao, J., et al., "Age of the fossil Dali Man in north-central China deduced from chronostratigraphy of the loess–paleosol sequence," *Quaternary Science Reviews* 21, no. 20–22 (2002): 2,191–98.

Zanolli, C., et al., "Denisovans: How to recognize them in the fossil record," 22nd Congress of the Indo-Pacific Prehistory Association (2022): 132–33.

14. Fossils Regained: Completing the Portrait

Bae, C. J., and X. Wu, "Making sense of eastern Asian Late Quaternary hominin variability," *Nature Communications* 15 (2024): 9479.

He, J., "Preliminary study on the teeth of Jinniushan archaic *Homo sapiens*," *Acta Anthropologica Sinica* 19, no. 3 (2000): 216–25.

Hublin, J., "Non, on ne vient pas de découvrir deux nouvelles espèces humaines!" *Pour la Science*, published online on August 18, 2021.

Ji, Q., et al., "Late Middle Pleistocene Harbin cranium represents a new *Homo* species," *The Innovation* 2, no. 3 (2021): 132.

Mbugua, M., "Ancient skeleton shines new light on evolution," *University of Delaware Messenger* 14, no. 2 (2006).

Ni, X., et al., "Massive cranium from Harbin in northeastern China establishes a new Middle Pleistocene human lineage," *The Innovation* 2, no. 3 (2021).

Rosenberg, K. R., et al., "Body size, body proportions, and encephalization in a Middle Pleistocene archaic human from northern China," *PNAS* 103, no. 10 (2006): 3,552–56.

"Significant Scientific Discovery: The First *Homo Heidelbergensis* Type Fossil in China," Hebei GEO University, published online on September 13, 2018.

Wei-Haas, M., "'Dragon Man' skull may be new species, shaking up human family tree," *National Geographic*, published online on June 25, 2021.

Wu, R., "The reconstruction of the fossil human skull from Jinniushan, Yinkou, Liaoning Province and its maintures," *Acta Anthropologica Sinica* 7, no. 2 (1988): 97–101.

Xia, H., et al., "Middle and Late Pleistocene Denisovan subsistence at Baishiya Karst Cave," *Nature* 632 (2024): 108–13.

Image Credits

p. viii: © Benoit CLARYS-Cité de la Préhistoire Aven d'Orgnac 12

p. 8: © Benoit CLARYS 24

p. 12: © Hendrik Schmidt/DPA/Photononstop

p. 14: Laurent Blondel © Albin Michel, "Hérédités génétiques nucléaire et mitochondriale"

p. 22: © Benoit CLARYS 24

p. 27: From Konopatskii, A. K. (ed. and trans. R. L. Bland and Y. V. Kuzmin), 2021. *Aleksei P. Okladnikov: The Great Explorer of the Past. Volume 2.* Archaeopress, Oxford. Reproduced with permission.

p. 30: © Dr. Bence Viola

p. 32: © Eva-Maria Geigl/CC BY-NC

p. 34: Laurent Blondel © Albin Michel, "Occupations humaines de la grotte de Denisova"

p. 40: © Benoit CLARYS-Cité de la Préhistoire Aven d'Orgnac 12

p. 43: Public domain

p. 45: Public domain

p. 46: © Henry Grant Collection/Mol/Shutterstock

p. 50: LVR-LandesMuseum

p. 51: *L'homme fossile de La Chapelle-aux-Saints*, Marcellin Boule/The Royal College of Surgeons of England

p. 53: The Field Museum, Image No. CSA66702

p. 58: © Benoit CLARYS 24

p. 61: Julia Margaret Cameron, restoration by Adam Cuerden

p. 63: Laurent Blondel © Albin Michel, "Évolution humaine des derniers 800,000 ans"

p. 66: Laurent Blondel © Albin Michel, "Flux géniques entre sapiens, néandertaliens, dénisoviens et super-archaïques"

p. 70: © Benoit CLARYS-SPM I Société suisse de Préhistoire et d'Archéologie 93

p. 72: The Filipino Negrito ethnic group known as Ayta Magbukon © ncip. gov.ph

p. 78: Laurent Blondel © Albin Michel, "Carte de l'héritage génétique dénisovien parmi les populations actuelles, et des principaux sites sapiens, néandertaliens et Denisova dans l'Altaï"

p. 82: © Benoit CLARYS-Editions Casterman 98

p. 84: © Shutterstock/beibaoke

p. 85: Public domain

p. 87: Public domain

p. 88: *The Skull of Sinanthropus pekinensis*, Franz Weidenreich

p. 91: Naturalis Biodiversity Center

p. 93: *Die Pithecanthropus-Schichten auf Java. Geologische und paläontologische Ergebnisse der Trinil-Expedition (1907 und 1908)*, Lenore Selenka and Max Blanckenhorn

p. 94: © akg-images/Interfoto

p. 102: © Photo12/Alamy/Marion Kaplan

p. 104: © Benoit CLARYS-Editions Casterman 98

p. 107: Laurent Blondel © Albin Michel, "Carte du plateau continental de la Sonde aux époques glaciaires et interglaciaires"

p. 112: Lee Roger Berger research team/Creative Commons Attribution 4.0 International

p. 115: Laurent Blondel © Albin Michel, "Carte des écozones climatiques de l'Ancien Monde"

p. 116: © Benoit CLARYS-Editions Casterman 98

p. 124: Laurent Blondel © Albin Michel, "Carte de la première sortie d'Afrique du genre Homo (avant 2 millions d'années–1.5 million d'années)"

p. 125: © Daniella Bar-Yosef

p. 127: Laurent Blondel © Albin Michel, "Carte de la deuxième sortie d'Afrique du genre Homo

(1.5 million d'années–800,000 ans)"

p. 134: © Benoit CLARYS 24

p. 136: Excavations of the Acheulian site of Gesher Benot Ya'aqov (Area B) © the GBY expedition

p. 143: Laurent Blondel © Albin Michel, "Évolution du volume endocrânien des principales forms humaines au cours des deux derniers millions d'années"

p. 145: Laurent Blondel © Albin Michel, "Carte de la troisième sortie d'Afrique du genre Homo

(800,000 ans–300,000 ans)"

p. 150: © Benoit CLARYS-Cedarc 98

p. 160: © Yamei Hou

p. 161: © Yamei Hou

p. 162: © Yamei Hou

p. 164: © Benoit CLARYS-Museum Velzeke

p. 166: © Shutterstock/Terri Price

p. 168: © Benoit CLARYS-Cité de la Préhistoire Aven d'Orgnac 12

p. 170: © Daniella Bar-Yosef

p. 175: Laurent Blondel © Albin Michel, "Carte de la Chine et emplacement des fossiles prédénisoviens et dénisoviens"

p. 178: © Benoit CLARYS 24

p. 182: Provided by Dongju Zhang from Lanzhou University

p. 184: Provided by Dongju Zhang from Lanzhou University

p. 185: © Liu Wu

p. 188: © Hill Debbie/UPI/ABACA

p. 189: Laurent Blondel © Albin Michel, "Phénotype du squelette dénisovien d'après le profil de methylation de Denisova 3"

p. 192: © Benoit CLARYS 24

p. 195: © Nian ZENG/GAMMA RAPHO

p. 196: © Liu Wu

p. 203: © Liu Wu

p. 205: © TaichungJohnny/Creative Commons Attribution 4.0 International

p. 208: © Benoit CLARYS 24

p. 210: Laurent Blondel © Albin Michel, "Les fragments du squelette de la femme de Jinniushan"

p. 220: © Benoit CLARYS 24

Insert

p. 1 (top): © Shutterstock/Igor Boshin

p. 1 (bottom): © Thilo Park/Wikimedia Commons

p. 2: © Science Photo Library/akg-images

p. 3 (top): © Smithsonian Digitization Program Office/Liang Bua Team

p. 3 (bottom): Public domain

p. 4 (top and bottom): Provided by Dongju Zhang from Lanzhou University

p. 5 (top and bottom): © Science Photo Library/Sucré Salé

p. 6 (top): © Didier Descouens/Creative Commons Attribution 4.0 International

p. 6 (bottom): © Liu Wu

p. 7: Xijun Ni et al., "Massive cranium from Harbin in northeastern China establishes a new Middle Pleistocene human lineage" © 2024 Elsevier B.V.

p. 8 (top): © Xiu-Jie Wu

p. 8 (bottom): Karen Rosenberg

Acknowledgments

Thank you to our editor, Christian Counillon, who has supported us for years in our efforts to introduce ancient humans to the general public. We also thank Xavier Müller and Alexandre Couëtoux for their help in developing and editing this text. We thank the wonderfully talented Benoit Clarys for the beautiful reconstructions of Denisova, which add so much to the book. We would also like to thank many colleagues and/or friends, in France and elsewhere, who generously answered our many questions about climate, fire, tools, fauna, the complex stratigraphic sections of certain sites, etc., and shared their invaluable field data and knowledge with us. Among them, we thank Gilles Ramstein of the UMR LSGE (Laboratory of Climate and Environmental Sciences); Jean-Jacques Bahain of the National Museum of Natural History; Jacques Chiaroni, Stéphane Mazières, and Jean-Luc Voisin of the UMR ADÉS at the University of Marseille; Clément Zanolli of the Pacea laboratory at the University of Bordeaux; Lutz Maul of the Senckenberg Research Institute in Frankfurt; Naama Goren and Liran Carmel of the Hebrew University of Jerusalem; Israel Hershkovitz and Daniella Bar-Yosef of Tel Aviv University; Ian Tattersall of the American Museum of Natural History in New York; Eric Delson at Lehman College of the City University of New York; Karen Rosenberg at the University of Delaware; and Liu Wu, Xiu-Jie Wu, and Yamei Hou of the Institute of Vertebrate Paleontology and Paleoanthropology of the Chinese Academy of Sciences. This major investigation of the Denisovan human is also the product of numerous exchanges and conversations with these distinguished researchers and other colleagues.

Index

About the Authors

SILVANA CONDEMI, a world-leading paleoanthropologist, is the research director of CNRS, the largest French public scientific research organization, at Aix-Marseille University. She is the coauthor of *A Pocket History of Human Evolution* and *The Secret World of Denisovans*.

FRANÇOIS SAVATIER is a journalist for the magazine *Pour la Science* (the French edition of *Scientific American*), where he focuses on the science of the past. He is the coauthor of *A Pocket History of Human Evolution* and *The Secret World of Denisovans*.